The Earth Clock I

Biological Nitrogen Fixation and the Evolution of Complex Life

2026 International

Expanded Edition

Qi Cheng

(Earth Clock Library Volume I)

Table of Content

Preface to the 2026 International Edition

Life on Earth is built upon a small number of fundamental processes that have shaped the history of our planet. Among them, biological nitrogen fixation stands as one of the most remarkable.

Every protein, every enzyme, every genome, and every living cell depends on nitrogen. Yet atmospheric nitrogen, despite comprising nearly eighty percent of the Earth's atmosphere, is inaccessible to most organisms. The emergence of biological nitrogen fixation therefore transformed a chemically limited world into one capable of supporting complex ecosystems and, ultimately, intelligent life.

For more than a century, scientists have investigated the microorganisms and enzymes responsible for nitrogen fixation. Their discoveries have revealed a fascinating biological system centered on nitrogenase, one of the most complex catalysts known in nature. Despite tremendous progress, the challenge of extending nitrogen-fixing capability to major crop plants remains one of the grand scientific ambitions of modern agriculture.

This book explores biological nitrogen fixation from a broader perspective. It examines not only the molecular mechanisms of nitrogenase, but also its evolutionary origins, its relationship with photosynthesis, and its significance in the development of life on Earth.

The framework presented here is based on the concept of the Earth Clock. In this view, Earth can be regarded as a giant biological clock whose history spans approximately ten billion years. Within this immense timescale, nitrogen fixation appears not as an isolated biochemical process, but as a critical event that influenced the trajectory of planetary evolution.

The future chapters also explore emerging ideas, including synthetic nitrogen-fixing organelles, light-utilising nitrogenases, artificial intelligence-guided enzyme design, and the possibility of engineering nitrogen fixation into future agricultural systems on Earth and beyond.

Science advances through curiosity, imagination and evidence. It is my hope that this book will encourage readers to view nitrogen fixation not merely as a topic in microbiology, but as a window into the past and future of life itself.

Qi Cheng

Cambridge, United Kingdom

2026

The Earth Clock I: Biological Nitrogen Fixation and the Evolution of Complex Life

Qi Cheng

Earth Clock Library Volume I

2026 International Expanded Edition

Chapter 1

The Earth Clock and Planetary Evolution

1.1 A Different Way to Read Earth's History

Human history occupies only a tiny fraction of the Earth's existence. Civilizations rise and fall within thousands of years, while biological evolution unfolds across millions of years and planetary evolution across billions.

To understand biological nitrogen fixation in its proper context, we must first step beyond the timescale of human experience and examine Earth itself as a dynamic system evolving through time.

The Earth Clock framework proposes a simple but powerful perspective:

> Earth can be viewed as a giant biological clock whose lifespan extends from its birth to its eventual destruction.

Under this framework, every geological event, evolutionary innovation, extinction, and biological breakthrough occupies a specific position on a planetary timeline.

The emergence of life, oxygenic photosynthesis, biological nitrogen fixation, multicellular organisms, flowering plants, animals, and human civilization are not isolated events. They represent sequential milestones in a much larger evolutionary journey.

1.2 The Birth of Earth

Current scientific evidence indicates that Earth formed approximately 4.54 billion years ago from the protoplanetary disk surrounding the young Sun.

During its earliest history, Earth was:

- Extremely hot
- Frequently bombarded by asteroids
- Geologically unstable

- Devoid of life

The primitive atmosphere likely contained:

- Carbon dioxide
- Water vapor
- Nitrogen
- Methane
- Hydrogen sulfide

Free oxygen was essentially absent.

For hundreds of millions of years, Earth remained a hostile world dominated by volcanic activity and planetary restructuring.

Yet within this seemingly chaotic environment, the conditions necessary for life gradually emerged.

1.3 The First Biological Revolution

At some point between 3.5 and 4.0 billion years ago, the first living systems appeared.

The exact mechanism remains unknown.

Several hypotheses have been proposed:

The Primordial Soup Model

Organic molecules accumulated in ancient oceans and gradually assembled into increasingly complex systems.

Hydrothermal Vent Hypothesis

Life originated near deep-sea hydrothermal systems where chemical energy was abundant.

RNA World Hypothesis

Self-replicating RNA molecules preceded modern DNA and proteins.

Panspermia

Life or its precursors arrived from extraterrestrial sources.

Regardless of its origin, life introduced an entirely new phenomenon into Earth's history:

> Information capable of self-replication and evolution.

The Earth Clock had begun to tick biologically.

1.4 The Nitrogen Barrier

Although early life emerged successfully, it immediately encountered a major limitation.

Nitrogen is essential for:

- Amino acids
- Proteins
- DNA
- RNA
- ATP
- Chlorophyll

Without biologically available nitrogen, life cannot expand.

Yet Earth's atmosphere consists largely of molecular nitrogen (N_2), whose triple bond makes it chemically inert.

$N \equiv N$

This bond is among the strongest in nature.

Most organisms cannot utilize atmospheric nitrogen directly.

As a result, early life faced a severe nitrogen limitation.

This challenge would shape the future evolution of the biosphere.

1.5 The Emergence of Nitrogen Fixation

One of the greatest innovations in the history of life was the evolution of biological nitrogen fixation. Certain ancient microorganisms developed the remarkable ability to convert atmospheric nitrogen into ammonia.

$$N_2+8H^++8e^-+16ATP\rightarrow 2NH_3+H_2+16ADP+16P_i$$

This reaction fundamentally changed the Earth.

For the first time, atmospheric nitrogen became biologically accessible.

The consequences were profound:

- Increased biomass
- Expansion of ecosystems
- Enhanced biological productivity
- Greater evolutionary diversification

Biological nitrogen fixation may therefore represent one of the earliest planetary-scale innovations.

1.6 Photosynthesis and the Second Revolution

If nitrogen fixation solved the nitrogen problem, photosynthesis solved the energy problem.

The emergence of photosynthetic organisms allowed life to harvest energy directly from sunlight.

Over time, oxygenic photosynthesis transformed Earth's atmosphere.

The Great Oxidation Event, approximately 2.4 billion years ago, dramatically increased atmospheric oxygen concentrations.

This event:

- Eliminated many anaerobic organisms
- Created opportunities for aerobic metabolism
- Enabled greater biological complexity
- Altered global geochemical cycles

Nitrogen fixation and photosynthesis became the two foundational engines of the biosphere.

1.7 The Earth Clock Model

The Earth Clock simplifies planetary history by scaling Earth's lifespan to approximately 10 billion years.

Within this framework:

Earth Clock Age	Major Event
0	Earth forms
10–15	First life
20–25	Nitrogen fixation emerges
25–30	Photosynthesis expands
45–50	Complex eukaryotes
60–70	Land plants
75–80	Dinosaurs
95–99	Humans
100	End of Earth

The model highlights a striking reality:

Human civilization occupies less than one percent of Earth's total biological history.

From the perspective of the Earth Clock, humanity has appeared only moments ago.

1.8 Humanity and Planetary Time

Humans often perceive themselves as central to Earth's story.

Yet the Earth Clock suggests a different perspective.

For billions of years:

- Microorganisms dominated the planet.
- Nitrogen-fixing bacteria maintained fertility.
- Photosynthetic organisms generated oxygen.
- Evolution proceeded without human intervention.

Modern civilization depends entirely upon biological systems established long before humans existed.

Agriculture, food production, atmospheric composition, and ecological stability all trace their origins to ancient microbial innovations.

In this sense, humanity is not outside the Earth Clock.

Human civilization is one of its latest products.

1.9 Looking Forward

Earth's history is unfinished.

Over the next billions of years:

- Solar luminosity will increase.
- Oceans may disappear.
- Atmospheric composition will change.
- Life will face new constraints.

At the same time, intelligent organisms may begin directing biological evolution through technology.

Artificial intelligence, synthetic biology, engineered nitrogen fixation, and future nitroplasts may represent entirely new evolutionary pathways.

For the first time in Earth's history, life may become capable of consciously modifying its own evolutionary future.

Whether humanity succeeds remains uncertain.

But one thing is clear:

The Earth Clock continues to tick.

And biological nitrogen fixation remains one of the most important mechanisms ever to emerge within it.

End of Chapter 1

Chapter 2

The Origin of Life

From Chemistry to Biology

2.1 The Greatest Transition in Earth History

Among all events that have occurred during the history of Earth, none is more profound than the origin of life.

Mountains rise and disappear.

Continents drift.

Oceans form and vanish.

Stars are born and die.

Yet life introduced something entirely new into the Universe:

> Matter capable of preserving information, reproducing itself, and evolving through time.

The emergence of life transformed Earth from a geologically active planet into a biologically active planet.

Without this transition, there would be:

- no ecosystems,

- no photosynthesis,
- no nitrogen fixation,
- no plants,
- no animals,
- no humans.

The Earth Clock therefore begins not with civilization, but with the first living systems.

2.2 The Early Earth Environment

Earth formed approximately 4.54 billion years ago.

The young planet was dramatically different from the modern world.

Characteristics included:

- intense volcanism,
- frequent asteroid impacts,
- high geothermal activity,
- reducing atmospheric conditions,
- absence of free oxygen.

The primitive atmosphere likely contained:

- carbon dioxide (CO_2),
- nitrogen (N_2),

- water vapor (H_2O),
- methane (CH_4),
- ammonia (NH_3),
- hydrogen sulfide (H_2S).

Oceans gradually formed as Earth cooled.

These oceans became the stage upon which the earliest biochemical processes unfolded.

Although conditions would appear hostile by modern standards, they may have provided exactly the environment required for life's emergence.

2.3 Defining Life

Before discussing the origin of life, we must first ask:

What is life?

There is no universally accepted definition.

However, most definitions include several common properties:

Organization

Living systems possess internal structure.

Metabolism

Living systems exchange energy and matter with their environment.

Replication

Living systems reproduce.

Evolution

Living systems change through natural selection.

Information Storage

Living systems preserve hereditary information.

Together, these properties distinguish living systems from nonliving matter.

The origin of life therefore represents the emergence of organized chemical systems capable of information-based evolution.

2.4 Chemical Evolution

One of the earliest scientific explanations for life's origin is the concept of chemical evolution.

This framework proposes that life emerged gradually through a series of increasingly complex chemical stages.

The process can be summarized as:

Stage 1

Formation of simple organic molecules.

Stage 2

Formation of biological macromolecules.

Stage 3

Assembly of molecular systems.

Stage 4

Emergence of self-replicating entities.

Stage 5

Evolution of primitive cells.

This sequence transformed chemistry into biology.

2.5 The Miller–Urey Experiment

In 1953, Stanley Miller and Harold Urey conducted one of the most famous experiments in biology.

They attempted to simulate early Earth conditions using:

- methane,
- ammonia,
- hydrogen,
- water vapor,

- electrical discharges.

After several days, amino acids appeared.

The experiment demonstrated that biologically relevant molecules could arise naturally from simple chemical precursors.

Although modern views of Earth's primitive atmosphere have evolved, the experiment remains historically significant.

It established the principle that organic molecules can emerge through natural processes.

2.6 Hydrothermal Vent Hypothesis

Another influential model proposes that life originated near hydrothermal vents.

These environments offer:

- abundant chemical energy,
- mineral catalysts,
- thermal gradients,
- protection from ultraviolet radiation.

Hydrothermal systems contain iron, sulfur, nickel and other metals that remain essential for modern biological processes.

Interestingly, many enzymes involved in ancient metabolism—including nitrogenase—utilize metal cofactors similar to those found in hydrothermal environments.

This observation has led some researchers to suggest that modern biological catalysts preserve chemical signatures from Earth's earliest ecosystems.

2.7 The RNA World

One of the most influential hypotheses in modern origin-of-life research is the RNA World hypothesis.

RNA possesses two remarkable properties:

1. It can store genetic information.
2. It can catalyze chemical reactions.

This dual functionality suggests that RNA may have preceded both DNA and proteins.

Under this model:

RNA → Protein Synthesis → DNA Genomes

The RNA World provides a plausible pathway connecting simple chemistry to modern cellular life. Although many questions remain unresolved, RNA continues to occupy a central position in origin-of-life research.

2.8 The First Cells

Eventually, molecular systems became enclosed within primitive membranes.

These structures created:

- internal environments,
- chemical boundaries,
- selective permeability.

The appearance of cellular organization represented a major evolutionary breakthrough.

Cells enabled:

- more efficient metabolism,
- improved replication,
- greater stability,
- long-term evolutionary innovation.

At this point, true biological evolution may have begun.

The Earth Clock had started its biological countdown.

2.9 The Last Universal Common Ancestor (LUCA)

All modern life shares remarkable biochemical similarities.

Examples include:

- DNA genomes,
- RNA transcription,
- protein synthesis,
- ATP metabolism,
- universal genetic code.

These shared features imply a common ancestry.

Scientists refer to this ancestral population as:

LUCA

(The Last Universal Common Ancestor)

LUCA was not the first living organism.

Rather, it represents the most recent common ancestor from which all modern life descended.

Although its exact characteristics remain uncertain, LUCA likely possessed:

- genetic information,
- protein synthesis machinery,
- membrane structures,
- energy metabolism.

Many researchers believe LUCA may already have possessed sophisticated biochemical systems.

2.10 Life and Nitrogen

The emergence of life immediately created a new challenge.

Living systems require nitrogen.

Nitrogen is essential for:

- amino acids,
- nucleotides,
- ATP,
- chlorophyll.

Yet atmospheric nitrogen is largely inaccessible.

This created one of the earliest resource limitations in Earth's history.

As biological populations expanded, demand for fixed nitrogen increased.

Eventually, evolutionary pressure favored organisms capable of utilizing atmospheric nitrogen directly.

This challenge set the stage for one of the most important innovations in Earth history:

Biological Nitrogen Fixation

The story of life therefore naturally leads to the story of nitrogen.

2.11 Was Nitrogen Fixation Inevitable?

The Earth Clock framework suggests a provocative possibility.

Given sufficient time, any biosphere based upon proteins and nucleic acids may eventually encounter a nitrogen limitation.

If atmospheric nitrogen is abundant but biologically inaccessible, evolution may repeatedly favor mechanisms capable of exploiting this vast reservoir.

From this perspective:

> Nitrogen fixation may represent not an accident of evolution, but an almost inevitable consequence of biological expansion.

If true, nitrogen fixation could be expected to emerge on many inhabited planets throughout the Universe.

This idea extends the significance of nitrogen fixation far beyond Earth.

2.12 The First Planetary Innovation

The origin of life transformed Earth from a chemical world into a biological world.

Yet life's continued expansion required further innovation.

The next great challenge was access to nitrogen.

The next great solution was nitrogen fixation.

The origin of life therefore represents only the first chapter of Earth's biological story.

The second chapter concerns how life learned to harvest one of the most abundant elements in the atmosphere.

That story begins with the evolution of nitrogen fixation.

Conclusions

The origin of life remains one of science's greatest mysteries.

Despite many unanswered questions, several conclusions are clear:

1. Life emerged early in Earth's history.
2. Biological evolution transformed the planet.
3. Cellular systems required continuous access to nitrogen.
4. Nitrogen availability became an important evolutionary constraint.
5. The emergence of nitrogen fixation would profoundly alter the future trajectory of life.

Within the Earth Clock framework, the origin of life marks the beginning of biological time.

The evolution of nitrogen fixation marks the beginning of biological expansion.

Together, they laid the foundation for everything that followed.

End of Chapter 2

Chapter 3

The Nitrogen Crisis of Early Earth

Nitrogen Limitation as a Planetary Bottleneck

3.1 A Planet Rich in Nitrogen but Poor in Life

One of the greatest paradoxes in Earth's history is that nitrogen has always been abundant.

Today, nitrogen gas constitutes approximately 78% of the atmosphere.

$$N_2\approx78\%\ of\ Earth's\ atmosphere$$

From a purely chemical perspective, Earth appears extraordinarily rich in nitrogen.

Yet early life experienced a very different reality.

Although nitrogen surrounded every organism, most of it was inaccessible.

The reason lies in the molecular structure of nitrogen gas.

Atmospheric nitrogen exists primarily as dinitrogen (N_2), stabilized by a triple bond:

$$N\equiv N$$

This bond possesses one of the highest dissociation energies found in nature.

Consequently, atmospheric nitrogen behaves almost like a locked reservoir.

The biosphere could see it, but could not easily use it.

3.2 The Nitrogen Problem

Every known form of life requires nitrogen.

Nitrogen forms essential components of:

- amino acids
- proteins
- nucleic acids
- ATP
- chlorophyll
- coenzymes

Without nitrogen:

- cells cannot grow,
- DNA cannot replicate,
- proteins cannot function.

As primitive organisms multiplied across early Earth, demand for fixed nitrogen increased steadily.

However, available nitrogen sources were limited.

This imbalance created what may be termed:

The Nitrogen Problem

The challenge was simple:

Life needed nitrogen faster than geology could supply it.

3.3 Sources of Fixed Nitrogen Before Biology

Before biological nitrogen fixation evolved, fixed nitrogen likely originated from several abiotic sources.

Lightning

Volcanism

Meteorite Impact

Hydrothermal Systems

↓

N_2 Activation

↓

NOx Formation

↓

Nitrate/Nitrite

↓

Prebiotic Chemistry

Lightning

Electrical discharges can convert atmospheric nitrogen into reactive nitrogen compounds.

Simplified reaction:

$N2+O2 \rightarrow 2NO$

$2NO+O2 \rightarrow 2NO2$

$3NO2+H2O \rightarrow 2HNO3+NO$

Overall,

$$N_2+O_2 \rightarrow NO_x$$

These compounds eventually entered soils and oceans.

Volcanic Activity

Volcanic emissions may have contributed small quantities of reduced nitrogen compounds.

Hydrothermal Systems

High-temperature geochemical reactions could generate limited bioavailable nitrogen.

Meteorites and Comets

Extraterrestrial materials may have delivered nitrogen-containing organic compounds.

Despite these inputs, global nitrogen availability remained relatively low.

Abiotic sources were insufficient to support large-scale biological expansion.

3.4 The First Global Resource Limitation

The Earth Clock Theory proposes that nitrogen availability represented one of the earliest planetary-scale biological bottlenecks.

A bottleneck occurs when a critical resource restricts system growth.

For early life:

- energy existed,
- water existed,
- carbon existed,

but usable nitrogen remained scarce.

As a result:

- biomass accumulation slowed,
- ecosystem expansion remained limited,
- evolutionary diversification was constrained.

Nitrogen therefore became one of the first major regulators of biosphere development.

3.5 The Nitrogen Bottleneck Hypothesis

The Nitrogen Bottleneck Hypothesis can be summarized as follows:

> Early biological evolution was constrained by limited access to fixed nitrogen.

Under this hypothesis:

Increasing nitrogen availability allows:

- larger populations,
- greater biodiversity,
- more complex ecosystems.

Decreasing nitrogen availability produces:

- ecological instability,
- reduced productivity,
- slower evolutionary innovation.

Nitrogen thus functions as a planetary control variable.

3.6 Nitrogen and Biological Information

One often overlooked consequence of nitrogen limitation concerns information storage.

Genes require nucleotides.

Nucleotides require nitrogen.

Therefore:

More nitrogen permits larger genomes.

Larger genomes enable:

- new proteins,
- regulatory networks,
- developmental complexity.

In this sense, nitrogen availability influences not only biomass but also biological information capacity.

The evolution of complexity may therefore depend partly upon the evolution of nitrogen acquisition systems.

3.7 The Ecological Consequences of Scarcity

Modern ecosystems provide insight into ancient conditions.

In many contemporary environments:

- deserts,
- oceans,
- grasslands,

nitrogen remains the primary limiting nutrient.

Even today, productivity frequently increases when nitrogen becomes available.

This suggests that nitrogen limitation has been a recurring feature throughout Earth's history.

Early Earth likely experienced similar pressures on a much larger scale.

3.8 Competition for Nitrogen

As biological populations expanded, competition intensified.

Organisms capable of obtaining nitrogen more efficiently gained evolutionary advantages.

Natural selection therefore favored:

- improved nitrogen uptake,
- nitrogen recycling,
- symbiotic associations,
- eventually nitrogen fixation itself.

The search for nitrogen became an evolutionary driver.

In many ways, the history of life can be viewed as a history of nutrient acquisition.

Among all nutrients, nitrogen occupied a uniquely important position.

3.9 Why Nitrogen Fixation Matters

Most evolutionary innovations improve survival.

Nitrogen fixation did something more.

It increased the carrying capacity of the entire planet.

Before nitrogen fixation:

Life depended on limited abiotic nitrogen sources.

After nitrogen fixation:

Life gained access to the largest nitrogen reservoir available:

the atmosphere.

This transformation fundamentally altered planetary ecology.

The biosphere no longer relied exclusively on geological processes.

Biology began generating its own nitrogen supply.

3.10 A Planetary Turning Point

Few biological innovations have changed Earth as profoundly as nitrogen fixation.

Its emergence likely resulted in:

- increased primary productivity,
- larger microbial populations,
- accelerated evolution,
- expanded food webs,
- greater ecosystem stability.

The Earth Clock Theory therefore regards nitrogen fixation as one of the most important transitions in Earth's history.

Not because it solved a local problem.

But because it solved a planetary problem.

3.11 Nitrogen Limitation Beyond Earth

The significance of nitrogen limitation may extend far beyond our planet.

Many exoplanets likely possess nitrogen-rich atmospheres.

If life exists elsewhere, similar challenges may arise.

The same physical laws apply:

- proteins require nitrogen,
- genomes require nitrogen,
- ecosystems require nitrogen.

Consequently:

Nitrogen acquisition mechanisms may represent universal evolutionary solutions.

Biological nitrogen fixation might not be unique to Earth.

It may be a predictable outcome of life itself.

3.12 Toward the Birth of Nitrogen Fixation

The nitrogen crisis of early Earth created immense evolutionary pressure.

Some organisms adapted.

Others disappeared.

Eventually, one lineage achieved a remarkable breakthrough:

the ability to reduce atmospheric nitrogen directly.

This innovation would reshape the planet.

It would influence the evolution of photosynthesis.

It would help support complex life.

It would become one of the oldest and most important biochemical processes on Earth.

The next chapter tells that story.

Conclusions

The Nitrogen Crisis of Early Earth represents a fundamental stage in planetary evolution.

Key conclusions include:

1. Atmospheric nitrogen was abundant but largely inaccessible.
2. Early life experienced chronic nitrogen limitation.
3. Abiotic nitrogen sources were insufficient for large-scale expansion.
4. Nitrogen availability acted as a planetary bottleneck.
5. Evolutionary pressure favored organisms capable of accessing atmospheric nitrogen.
6. The emergence of biological nitrogen fixation resolved one of the earliest resource crises in Earth's history.

Within the Earth Clock framework, the nitrogen crisis represents the first major challenge faced by the biosphere.

The evolution of nitrogen fixation represents the first major solution.

End of Chapter 3

Chapter 4

The Birth of Nitrogen Fixation

The First Great Biological Innovation

4.1 Introduction

Life began as a chemical phenomenon.

Nitrogen fixation transformed it into a planetary phenomenon.

Among all biochemical innovations that emerged during the first half of Earth's history, few rival the significance of biological nitrogen fixation.

The evolution of nitrogenase provided life with direct access to the largest nitrogen reservoir on the planet:

the atmosphere.

This breakthrough fundamentally altered Earth's biological future.

Within the Earth Clock framework, the emergence of nitrogen fixation represents:

The First Great Biological Innovation.

Before nitrogen fixation, life depended largely on external nitrogen inputs.

After nitrogen fixation, life began generating its own supply.

This transition changed the scale of biological evolution forever.

4.2 The Ancient Earth

To understand the birth of nitrogen fixation, we must first revisit early Earth.

Approximately 4.0–3.5 billion years ago:

- oxygen was scarce,
- oceans dominated the planet,
- microorganisms ruled the biosphere,
- atmospheric nitrogen was abundant.

The microbial world was vastly different from modern ecosystems.

Most organisms were:

- anaerobic,
- unicellular,
- metabolically simple.

Yet even these primitive organisms required nitrogen.

Competition for nitrogen intensified as populations expanded.

Evolution began searching for solutions.

4.3 The Evolutionary Challenge

The central challenge facing early life was straightforward:

How can a cell access atmospheric nitrogen?

The difficulty lay in the strength of the nitrogen triple bond.

$N\equiv N$

Breaking this bond requires substantial energy.

Few chemical systems are capable of performing this reaction efficiently.

Any organism capable of reducing atmospheric nitrogen would immediately gain a major competitive advantage.

Natural selection therefore strongly favored mechanisms capable of exploiting this untapped resource.

4.4 The Emergence of Nitrogenase

At some point during early evolution, a remarkable catalytic system emerged:

Nitrogenase

Nitrogenase remains the only known biological enzyme capable of reducing atmospheric nitrogen directly.

The reaction it catalyzes is:

$$N_2+8H^++8e^-+16ATP\rightarrow 2NH_3+H_2+16ADP+16P_i$$

This reaction remains one of the most complex transformations in biology.

Nitrogenase operates under ambient temperature and pressure conditions that industrial chemistry struggles to match efficiently.

The emergence of nitrogenase therefore represents an extraordinary evolutionary achievement.

4.5 Did Nitrogen Fixation Precede Photosynthesis?

One of the most important questions in evolutionary biology concerns timing.

Which came first:

- nitrogen fixation,
- or photosynthesis?

The answer remains uncertain.

However, several observations suggest that nitrogen fixation may be extremely ancient.

Evidence 1

Nitrogenase is widespread among deeply branching microbial lineages.

Evidence 2

Nitrogenase-like proteins appear throughout diverse evolutionary groups.

Evidence 3

The oxygen sensitivity of nitrogenase suggests evolution in an anaerobic world.

This observation is particularly important.

Modern oxygen-rich conditions are hostile to nitrogenase.

Therefore the enzyme likely originated before major oxygen accumulation.

This places nitrogen fixation very early in Earth history.

4.6 LUCA and Nitrogen Fixation

The Last Universal Common Ancestor (LUCA) remains one of biology's greatest mysteries.

Scientists continue debating whether LUCA possessed nitrogen-fixing capability.

Two possibilities exist.

Scenario A

LUCA possessed nitrogen fixation.

Under this model:

Nitrogen fixation predates the divergence of Bacteria and Archaea.

This would make nitrogen fixation one of the oldest metabolic innovations on Earth.

Scenario B

Nitrogen fixation evolved shortly after LUCA.

In this case:

Multiple microbial lineages subsequently inherited the capability through vertical inheritance and horizontal gene transfer.

Both scenarios imply extreme antiquity.

Either way, nitrogen fixation appears closely linked to the earliest phases of biological evolution.

4.7 The Metal World

Nitrogenase is fundamentally a metal-dependent enzyme.

Three major forms are known today.

Molybdenum Nitrogenase

The most efficient and widespread system.

Contains:

- molybdenum,
- iron,
- sulfur.

Key catalytic center:

FeMoco

(iron–molybdenum cofactor)

Vanadium Nitrogenase

Contains:

- vanadium,
- iron,
- sulfur.

Key catalytic center:

FeVco

(iron–vanadium cofactor)

Iron-Only Nitrogenase

Contains:

- iron,
- sulfur.

Functions under specific environmental conditions.

The existence of multiple nitrogenase systems suggests a long evolutionary history influenced by changing planetary geochemistry.

4.8 The Geochemical Connection

Earth's crust and oceans changed dramatically during early evolution.

Metal availability fluctuated through time.

Different environments favored different catalytic strategies.

The Earth Clock Theory proposes that nitrogenase evolution was closely coupled to planetary geochemistry.

As Earth evolved:

- metal availability changed,
- redox conditions changed,
- biological opportunities changed.

Nitrogenase evolution reflects these transitions.

In this sense, nitrogenase serves as a molecular fossil preserving information about ancient Earth.

4.9 Horizontal Gene Transfer

One reason nitrogen fixation spread so widely is horizontal gene transfer.

Unlike many biological traits, nitrogen fixation genes can move between unrelated organisms.

This process accelerated the expansion of nitrogen-fixing capability.

As nitrogenase spread:

- ecosystems became more productive,
- nitrogen cycling intensified,
- biological diversity increased.

Nitrogen fixation was no longer restricted to a single lineage.

It became a planetary phenomenon.

4.10 The First Global Nitrogen Cycle

Before nitrogen fixation, nitrogen cycling was largely geochemical.

After nitrogen fixation evolved:

biology entered the cycle.

The modern nitrogen cycle emerged gradually.

Key processes include:

- nitrogen fixation,
- ammonification,
- nitrification,
- denitrification.

Together these processes regulate the movement of nitrogen through:

- atmosphere,
- oceans,
- soils,
- organisms.

Nitrogen fixation became the gateway through which atmospheric nitrogen entered biological systems.

4.11 Why Nitrogen Fixation Changed Everything

Many biological innovations improve individual fitness.

Nitrogen fixation improved planetary fitness.

The consequences included:

Increased Biomass

More nitrogen enabled greater productivity.

Larger Ecosystems

Food webs expanded.

Greater Biodiversity

New ecological niches appeared.

Accelerated Evolution

Larger populations generated more variation.

Increased Biological Complexity

More fixed nitrogen supported larger genomes and more sophisticated organisms.

The emergence of nitrogen fixation therefore influenced not only individual organisms but the entire trajectory of life.

4.12 The First Great Biological Innovation

Within the Earth Clock framework, biological innovations can be ranked according to their planetary impact.

Among the earliest innovations:

1. Origin of Life
2. Nitrogen Fixation
3. Photosynthesis
4. Aerobic Respiration
5. Multicellularity

Nitrogen fixation occupies a unique position.

The origin of life created biology.

Nitrogen fixation enabled biological expansion.

Photosynthesis later transformed biological energy.

This sequence suggests that nitrogen fixation was one of the earliest mechanisms permitting large-scale biosphere growth.

4.13 Nitrogen Fixation Beyond Earth

If life exists elsewhere in the Universe, similar challenges may arise.

Any biosphere based upon:

- proteins,
- nucleic acids,
- cellular metabolism,

requires nitrogen.

Consequently, nitrogen acquisition systems may evolve repeatedly.

The Earth Clock Theory therefore predicts:

> Nitrogen fixation or functionally equivalent processes may be common among inhabited planets.

This idea links microbiology to astrobiology.

4.14 Toward the Oxygen Revolution

Nitrogen fixation solved one of the earliest nutrient crises.

However, another challenge remained:

energy.

The biosphere still depended upon limited chemical energy sources.

A second great innovation would soon emerge.

That innovation was photosynthesis.

The interaction between nitrogen fixation and photosynthesis would ultimately transform the planet.

The next chapters explore this remarkable relationship.

Conclusions

The birth of nitrogen fixation represents one of the most important events in Earth's history.

Key conclusions include:

1. Nitrogen fixation likely evolved very early.
2. Nitrogenase remains one of the most sophisticated enzymes in biology.
3. Multiple nitrogenase systems reflect ancient evolutionary diversification.
4. Nitrogen fixation transformed Earth's nitrogen economy.
5. The expansion of biological productivity depended heavily upon fixed nitrogen availability.
6. Nitrogen fixation may represent the first great biological innovation after the origin of life.

Within the Earth Clock framework, nitrogen fixation marks the moment when life gained direct access to the atmosphere.

This breakthrough allowed biology to expand from a limited chemical phenomenon into a planetary force.

End of Chapter 4

Chapter 5

Nitrogenase

The Enzyme That Changed Earth

5.1 Introduction

Throughout Earth's history, countless enzymes have evolved to catalyze specific biochemical reactions.

Some digest food.

Some synthesize DNA.

Others generate energy.

Yet among all known biological catalysts, few rival the importance of nitrogenase.

Nitrogenase performs a reaction that fundamentally altered the trajectory of life on Earth:

the conversion of atmospheric nitrogen into biologically available nitrogen.

Without nitrogenase:

- ecosystems would be smaller,
- biodiversity would be lower,
- agriculture would not exist in its current form,
- complex life might never have evolved.

For this reason, nitrogenase may be regarded as:

The Enzyme That Changed Earth.

5.2 Discovery of Nitrogenase

Although farmers had recognized the benefits of leguminous crops for thousands of years, the biochemical basis remained mysterious.

During the nineteenth and twentieth centuries, scientists gradually discovered:

- symbiotic nitrogen fixation,
- free-living nitrogen-fixing bacteria,
- biological nitrogen fixation mechanisms.

These discoveries culminated in the identification of nitrogenase as the enzyme responsible for atmospheric nitrogen reduction.

The revelation transformed nitrogen fixation from an agricultural observation into a molecular science.

5.3 A Remarkable Chemical Reaction

The reaction catalyzed by nitrogenase appears deceptively simple.

Atmospheric nitrogen is converted into ammonia.
However, the chemistry is extraordinarily challenging.

The overall reaction is:

$N_2+8H^++8e^-+16ATP\rightarrow2NH_3+H_2+16ADP+16P_i$

This transformation involves:

- breaking the N≡N triple bond,
- transferring multiple electrons,
- coordinating proton transfer,
- coupling ATP hydrolysis to catalysis.

Few reactions in biology require such sophisticated molecular machinery.

5.4 The Architecture of Nitrogenase

The classical molybdenum nitrogenase consists of two major components:

Component I

MoFe Protein

Also called:

Dinitrogenase

This is the catalytic core.

Functions:

- binds substrate,
- contains catalytic cofactors,
- performs nitrogen reduction.

Component II

Fe Protein

Also called:

Dinitrogenase reductase

Functions:

- transfers electrons,
- hydrolyzes ATP,
- powers catalysis.

Together these proteins form one of nature's most elegant molecular systems.

5.5 The Fe Protein

The Fe protein is relatively small but critically important.

Its key features include:

- homodimeric structure,
- [4Fe–4S] cluster,
- ATP-binding sites.

The Fe protein acts as an electron delivery system.

Each catalytic cycle involves:

1. ATP binding
2. Electron transfer
3. Association with MoFe protein
4. ATP hydrolysis
5. Dissociation

This process repeats many times during nitrogen reduction.

The Fe protein functions almost like a molecular battery charger.

5.6 The MoFe Protein

The MoFe protein is the catalytic heart of nitrogenase.

It contains two remarkable metal clusters:

P Cluster

Acts as an intermediate electron relay.

FeMoco

The catalytic center.

FeMoco consists of:

- iron,
- sulfur,
- molybdenum,
- carbon.

For decades, FeMoco represented one of the greatest structural mysteries in biology.

Its complex architecture remains among the most sophisticated metal clusters known.

5.7 FeMoco:

Nature's Masterpiece

The iron–molybdenum cofactor (FeMoco) is often considered one of the most extraordinary structures in biochemistry.

Its architecture allows:

- electron accumulation,
- substrate binding,
- bond activation,

- nitrogen reduction.

For many researchers, FeMoco represents an evolutionary masterpiece.

The existence of such a complex structure raises profound questions:

How did evolution assemble it?

Why did it emerge so early?

What alternative forms may still exist in nature?

These questions continue to inspire research worldwide.

5.8 The Catalytic Cycle

Nitrogenase operates through a series of electron transfer events.

The general sequence is:

Step 1

Electron enters Fe protein.

Step 2

ATP binds Fe protein.

Step 3

Fe protein associates with MoFe protein.

Step 4

Electron transferred to P cluster.

Step 5

Electron transferred to FeMoco.

Step 6

Nitrogen reduction progresses.

Step 7

ATP hydrolyzed.

Step 8

Complex dissociates.

The cycle repeats multiple times until nitrogen reduction is complete.

This mechanism requires exquisite molecular coordination.

5.9 Why So Much ATP?

One of the most puzzling aspects of nitrogenase is its enormous energy requirement.

At least sixteen ATP molecules are consumed for each nitrogen molecule reduced.

Why?

The answer lies in the extraordinary stability of N_2. The nitrogen triple bond evolved as a planetary reservoir because it is so difficult to break.

Nitrogenase solves this challenge by coupling:

- ATP hydrolysis,
- electron transfer,
- metal-cluster chemistry.

The process is energetically expensive but biologically effective.

5.10 Oxygen:
Nitrogenase's Greatest Enemy

Despite its remarkable catalytic power, nitrogenase possesses a major weakness:

oxygen sensitivity.

Exposure to oxygen can damage:

- Fe protein,
- FeMoco,
- electron transfer pathways.

This vulnerability likely reflects the enzyme's ancient origin.

Nitrogenase evolved in an anaerobic world long before atmospheric oxygen accumulated.

As oxygen increased, organisms developed protective strategies:

- heterocysts,
- root nodules,
- respiratory protection,
- temporal separation.

These adaptations enabled nitrogen fixation to persist in an oxygen-rich biosphere.

5.11 Alternative Nitrogenases

Not all nitrogenases contain molybdenum.

Two alternative systems are known.

Vanadium Nitrogenase

Contains:

- vanadium,
- iron,
- sulfur.

Functions under molybdenum limitation.

Iron-Only Nitrogenase

Contains:

- iron,
- sulfur.

Represents a more simplified catalytic system.

The existence of multiple nitrogenases suggests a long evolutionary history shaped by changing environmental conditions.

5.12 Nitrogenase and Evolution

Nitrogenase is more than an enzyme.

It is an evolutionary archive.

Its structure preserves information about:

- ancient oceans,
- metal availability,
- atmospheric conditions,
- microbial evolution.

Many researchers believe nitrogenase originated before major oxygen accumulation.

If true, nitrogenase may be among the oldest enzymes still functioning today.

Few molecular systems provide such a direct connection to Earth's distant past.

5.13 Nitrogenase and the Earth Clock

Within the Earth Clock framework, nitrogenase occupies a unique position.

The origin of life created biological systems.

Nitrogenase enabled their expansion.

Photosynthesis later transformed planetary energy flow.

Together these innovations shaped Earth's biological future.

The Earth Clock therefore recognizes nitrogenase as one of the foundational molecular inventions of life.

Without nitrogenase, Earth's biosphere would likely look dramatically different.

5.14 Nitrogenase and Future Agriculture

Human civilization depends heavily on nitrogen.

Modern agriculture relies largely on industrial fertilizer production.

Yet biological nitrogen fixation remains more sustainable.

Future strategies may include:

- enhanced symbioses,
- engineered cereals,
- synthetic nitroplasts,
- artificial nitrogen-fixing organelles,
- Light-Utilising Nitrogenase (LUN).

All of these approaches build upon principles first established by nitrogenase billions of years ago.

5.15 Nitrogenase Beyond Earth

The significance of nitrogenase may eventually extend beyond Earth.

Future:

- lunar habitats,
- Martian settlements,

- deep-space ecosystems,

will require efficient nitrogen cycling.

Nitrogenase or nitrogenase-inspired systems may become essential technologies for sustaining life beyond our planet.

The enzyme that helped transform Earth may one day help humanity inhabit new worlds.

Conclusions

Nitrogenase is one of the most remarkable enzymes ever discovered.

Key conclusions include:

1. Nitrogenase catalyzes the biological reduction of atmospheric nitrogen.
2. The enzyme consists of Fe protein and MoFe protein.
3. FeMoco is among the most sophisticated catalytic centers known.
4. Nitrogenase likely originated very early in Earth's history.

5. Oxygen sensitivity reflects its ancient evolutionary origin.
6. Nitrogenase transformed planetary nitrogen availability.
7. Future biological engineering efforts continue to draw inspiration from nitrogenase.

Within the Earth Clock framework, nitrogenase represents far more than a biochemical catalyst.

It is a molecular invention that helped shape the history of life itself.

Life began with chemistry.

Nitrogenase helped turn chemistry into a biosphere.

End of Chapter 5

Chapter 6

Evolution of Nitrogenase Systems

From Ancient Metal Catalysts to Modern Nitrogen Fixation

6.1 Introduction

Nitrogenase is often described as a single enzyme.

In reality, modern biology recognizes a family of related nitrogen-fixing systems that have diversified over billions of years.

Today three major classes are known:

Molybdenum Nitrogenase

(nif system)

Vanadium Nitrogenase

(vnf system)

Iron-Only Nitrogenase

(anf system)

Each system performs the same essential function: the reduction of atmospheric nitrogen.

Yet they differ significantly in:

- metal composition,

- catalytic efficiency,
- environmental distribution,
- evolutionary history.

The existence of multiple nitrogenases provides important clues about the earliest stages of biological evolution.

Within the Earth Clock framework, these systems may be viewed as molecular fossils recording the changing chemistry of ancient Earth.

6.2 The Nitrogenase Family

Modern nitrogenases belong to a broader superfamily of metalloproteins.

Despite their differences, all nitrogenases share several common characteristics:

Common Features

- ATP-dependent catalysis
- Metal–sulfur clusters
- Electron transfer pathways
- Homologous structural architecture
- Conserved evolutionary origins

These similarities strongly suggest that all nitrogenases evolved from a common ancestral system.

The challenge is determining which form appeared first.

6.3 The Three Nitrogenase Systems

6.3.1 Molybdenum Nitrogenase (nif)

The most widespread and efficient nitrogenase.

Catalytic center:

FeMoco

Contains:

- molybdenum
- iron
- sulfur
- carbon

Characteristics:

- highest catalytic efficiency
- broadest distribution
- dominant in modern ecosystems

For many years, molybdenum nitrogenase was assumed to represent the original nitrogen-fixing system.

However, recent evidence suggests a more complicated history.

6.3.2 Vanadium Nitrogenase (vnf)

Vanadium nitrogenase functions when molybdenum availability is limited.

Catalytic center:

FeVco

Contains:

- vanadium
- iron
- sulfur

Characteristics:

- lower efficiency
- distinct substrate specificity
- broader catalytic versatility

Interestingly, vanadium nitrogenase can reduce certain substrates more efficiently than molybdenum nitrogenase.

This suggests that alternative nitrogenases may preserve ancient catalytic strategies.

6.3.3 Iron-Only Nitrogenase (anf)

The simplest known nitrogenase.

Contains:

- iron
- sulfur

No molybdenum.

No vanadium.

Characteristics:

- lowest efficiency
- restricted distribution
- potentially ancient evolutionary features

Its simplicity has led many researchers to propose that iron-only systems may resemble ancestral nitrogenases.

6.4 The Metal Age Hypothesis

The Earth Clock Theory introduces a concept called:

The Metal Age Hypothesis

This hypothesis proposes that nitrogenase evolution was strongly influenced by the availability of transition metals in ancient oceans.

As Earth's chemistry changed, different metals became available.

Consequently, different nitrogenase systems emerged.

The sequence may have been linked directly to planetary geochemistry.

6.5 Ancient Oceans and Metal Availability

The chemistry of modern oceans differs dramatically from that of early Earth.

Ancient oceans were:

- oxygen-poor
- iron-rich
- chemically reducing

Large quantities of dissolved iron were available.

By contrast:

molybdenum concentrations were much lower.

Vanadium availability fluctuated through time.

These conditions may have shaped early nitrogenase evolution.

6.6 Which Nitrogenase Came First?

One of the most important unresolved questions is:

Which nitrogenase evolved first?

Several hypotheses have been proposed.

Hypothesis A

Mo Nitrogenase First

Traditional model.

Advantages:

- highest efficiency
- dominant modern distribution

Problem:

Molybdenum may have been scarce on early Earth.

Hypothesis B

V Nitrogenase First

Intermediate model.

Advantages:

- compatible with some ancient geochemical scenarios

Problem:

Evidence remains limited.

Hypothesis C

Fe-Only Nitrogenase First

Increasingly discussed model.

Advantages:

- iron abundance on early Earth
- structural simplicity
- geochemical plausibility

Under this model:

Fe → V → Mo

represents a progressive optimization process.

The Earth Clock framework considers this possibility particularly attractive.

6.7 Evolution Through Optimization

Natural selection often favors increasing efficiency.

Under the Metal Age Hypothesis:

Stage 1

Iron-only nitrogenase

Simple but inefficient.

Stage 2

Vanadium nitrogenase

Improved catalytic flexibility.

Stage 3

Molybdenum nitrogenase

Maximum efficiency.

This sequence resembles many technological innovations.

Early solutions function.

Later solutions improve.

Evolution behaves similarly.

Nitrogenase may represent billions of years of catalytic optimization.

6.8 Nitrogenase and Hydrothermal Vents

Many origin-of-life models emphasize hydrothermal systems.

These environments contain:

- iron
- sulfur
- nickel
- molybdenum

all important components of ancient metabolism.

Some researchers propose that nitrogenase evolved within hydrothermal ecosystems.

Several observations support this possibility:

Metal-rich environments

Hydrothermal systems supply catalytic metals.

Ancient microbial communities

Modern vent ecosystems host deeply branching lineages.

Anaerobic conditions

Consistent with nitrogenase oxygen sensitivity.

Although direct evidence remains unavailable, hydrothermal origins remain plausible.

6.9 Horizontal Gene Transfer and Global Expansion

Nitrogen fixation genes spread extensively through horizontal gene transfer.

Unlike many biological traits, nif genes can move between unrelated lineages.

Consequences include:

- rapid ecological expansion
- increased nitrogen fixation capacity
- diversification of microbial communities

Horizontal gene transfer helped transform nitrogen fixation from a local innovation into a planetary process.

This expansion dramatically altered Earth's biosphere.

6.10 Nitrogenase Family Tree

Modern phylogenetic analyses suggest a complex evolutionary history.

A simplified model appears below.

```
Ancestral Nitrogenase

            |

     ----------------

     |                  |

 Early Variants

     |

 ------------------------

 |              |              |

 anf            vnf            nif
(Fe)            (V)            (Mo)
```

Although details remain debated, all major nitrogenases likely share a common origin.

The family tree reflects billions of years of evolutionary experimentation.

6.11 Nitrogenase-Like Systems

One of the most exciting developments in recent decades has been the discovery of nitrogenase-like enzymes.

Examples include:

DPOR

Dark-operative protochlorophyllide oxidoreductase

COR

Chlorophyllide oxidoreductase

Other Nitrogenase-Like Systems

Involved in:

- cofactor biosynthesis
- tetrapyrrole metabolism
- specialized catalytic reactions

These systems suggest that nitrogenase evolution generated an entire catalytic superfamily.

Nitrogen fixation may therefore represent only one branch of a much larger evolutionary story.

6.12 Evolution and Planetary Change

Nitrogenase evolution cannot be separated from planetary evolution.

As Earth changed:

- oceans changed,
- atmosphere changed,
- metal availability changed,
- ecological opportunities changed.

Nitrogenase adapted accordingly.

This close relationship illustrates a central principle of the Earth Clock Theory:

> Biological evolution and planetary evolution are inseparable processes.

Organisms evolve within planets.

Planets shape evolutionary possibilities.

6.13 Nitrogenase as a Molecular Fossil

Many ancient organisms have disappeared.

Ancient oceans no longer exist.

Ancient atmospheres are gone.

Yet nitrogenase survives.

Its structure preserves information about:

- ancient chemistry,
- ancient environments,
- ancient ecosystems.

Nitrogenase therefore functions as a molecular fossil. By studying it, scientists gain insight into some of the earliest chapters of life's history.

6.14 Toward Nitrogenase Diversity

Evolution rarely stops.

New nitrogenase-like systems may still await discovery.

Future research may reveal:

- oxygen-tolerant variants,
- photochemical variants,
- novel metal cofactors,
- entirely new catalytic architectures.

The Earth Clock framework predicts that nitrogenase evolution is not finished.

The future may produce systems as different from modern nitrogenase as modern nitrogenase is from its earliest ancestors.

Conclusions

The evolution of nitrogenase systems reflects the intertwined history of life and Earth.

Key conclusions include:

1. Three major nitrogenase systems exist today: nif, vnf and anf.
2. Metal availability likely influenced nitrogenase evolution.
3. Iron-only nitrogenase may preserve ancient characteristics.
4. Horizontal gene transfer accelerated the spread of nitrogen fixation.
5. Nitrogenase belongs to a larger family of related enzymes.
6. Nitrogenase evolution mirrors changes in planetary geochemistry.

7. Nitrogenase serves as a molecular fossil recording Earth's biological history.

Within the Earth Clock framework, nitrogenase evolution represents one of the clearest examples of how life adapts to planetary change.

The story of nitrogenase is therefore not merely a biological story.

It is also the story of Earth itself.

End of Chapter 6

Chapter 7

Alternative Nitrogenases

Evolutionary Experiments in Biological Nitrogen Fixation

7.1 Introduction

For many decades, biological nitrogen fixation was thought to depend exclusively upon molybdenum nitrogenase.

The discovery of alternative nitrogenases fundamentally changed this view.

Scientists eventually identified two additional nitrogen-fixing systems:

Vanadium Nitrogenase

(vnf system)

Iron-Only Nitrogenase

(anf system)

These discoveries revealed that nature has evolved multiple solutions to the same biochemical challenge.

The existence of alternative nitrogenases demonstrates a fundamental principle of evolution:

There is rarely only one way to solve an important biological problem.

Within the Earth Clock framework, alternative nitrogenases may represent evolutionary experiments spanning billions of years.

7.2 Why Alternative Nitrogenases Exist

At first glance, alternative nitrogenases appear redundant.

If molybdenum nitrogenase is more efficient, why would other systems evolve?

The answer lies in environmental variability.

Throughout Earth's history:

- metal availability fluctuated,
- ocean chemistry changed,
- ecosystems diversified.

Under some conditions:

molybdenum becomes scarce.

Organisms possessing alternative catalytic systems gain a selective advantage.

Alternative nitrogenases therefore function as evolutionary backup systems.

7.3 Vanadium Nitrogenase

Vanadium nitrogenase was the first alternative nitrogenase discovered.

Its catalytic center contains:

- vanadium
- iron
- sulfur

instead of molybdenum.

The overall reaction remains similar:

Nitrogen is reduced to ammonia.

However, important differences exist.

Structural Similarity

Vanadium nitrogenase closely resembles molybdenum nitrogenase.

Catalytic Differences

Reaction rates differ.

Substrate preferences differ.

By-product formation differs.

These differences provide valuable insights into nitrogenase evolution.

7.4 Iron-Only Nitrogenase

Iron-only nitrogenase represents the simplest known nitrogen-fixing system.

Its catalytic center lacks both:

- molybdenum
- vanadium

and relies entirely upon iron-based chemistry.

This simplicity has attracted considerable attention.

Some researchers suggest that iron-only nitrogenase may preserve ancient characteristics inherited from early Earth environments.

If correct, iron-only nitrogenase could provide a glimpse into some of the earliest stages of biological nitrogen fixation.

7.5 Ancient Oceans and Alternative Nitrogenases

The Earth Clock Theory places alternative nitrogenases within a broader geological context.

Ancient oceans differed dramatically from modern oceans.

Conditions included:

- low oxygen,
- abundant dissolved iron,
- fluctuating trace metals.

Under such conditions:

iron-based systems may have been particularly advantageous.

As planetary chemistry changed, nitrogenase systems diversified.

Alternative nitrogenases therefore record environmental transitions preserved across geological time.

7.6 Catalytic Diversity

One of the most fascinating aspects of alternative nitrogenases is their catalytic flexibility.

Classical nitrogenase primarily reduces nitrogen.

Alternative nitrogenases display broader substrate utilization.

Examples include:

- acetylene reduction,
- carbon monoxide reduction,
- hydrocarbon production.

These activities reveal unexpected catalytic capabilities.

Nitrogen fixation may represent only one aspect of a larger catalytic potential.

7.7 Carbon Monoxide Reduction

A particularly important discovery was the ability of some alternative nitrogenases to reduce carbon monoxide.

Simplified representation:

CO\rightarrow Hydrocarbons

This observation surprised many researchers.

Carbon monoxide reduction resembles aspects of industrial catalytic chemistry.

The discovery suggested that nitrogenases possess remarkable catalytic versatility.

This finding broadened scientific interest beyond nitrogen fixation alone.

7.8 Nitrogenases and Hydrocarbon Formation

Some alternative nitrogenases can generate small hydrocarbons under experimental conditions.

Potential products include:

- methane
- ethylene
- ethane

Although efficiencies remain low, these observations have important implications.

They demonstrate that nitrogenase chemistry extends beyond nitrogen metabolism.

The enzyme family may possess latent catalytic capabilities that evolution has only partially explored.

7.9 Evolutionary Significance

Why would nature maintain multiple nitrogenase systems?

Several explanations are possible.

Environmental Adaptation

Different systems function under different geochemical conditions.

Resource Flexibility

Alternative metals provide resilience.

Catalytic Innovation

Alternative systems enable exploration of new reactions.

Evolutionary Insurance

Multiple systems reduce vulnerability to environmental change.

Together these factors enhance long-term evolutionary success.

7.10 Alternative Nitrogenases as Evolutionary Archives

The Earth Clock Theory proposes that alternative nitrogenases preserve information about earlier evolutionary stages.

Each system reflects:

- metal availability,
- environmental conditions,
- ecological pressures.

They are not merely alternative enzymes.

They are records of evolutionary history.

Studying them helps reconstruct ancient Earth environments.

7.11 The Hidden Diversity of Nitrogenase

Current nitrogenases may represent only a fraction of existing diversity.

Many microbial lineages remain poorly characterized.

Future discoveries may reveal:

- novel metal cofactors,
- hybrid systems,
- oxygen-tolerant variants,

- photochemical variants.

The evolutionary story of nitrogenase is likely incomplete.

Nature may still contain catalytic strategies not yet imagined.

7.12 Alternative Nitrogenases and Synthetic Biology

The catalytic flexibility of alternative nitrogenases makes them attractive targets for engineering.

Potential applications include:

Sustainable Agriculture

Improved biological nitrogen fixation.

Carbon Recycling

Carbon monoxide conversion.

Renewable Chemicals

Biological hydrocarbon production.

Artificial Metabolism

Novel synthetic pathways.

Alternative nitrogenases may therefore contribute to future biotechnology as well as evolutionary research.

7.13 Implications for the LUN Hypothesis

The existence of multiple nitrogenase systems supports an important conclusion:

Nitrogenase evolution is highly adaptable.

If evolution produced:

- Mo nitrogenase,
- V nitrogenase,
- Fe-only nitrogenase,

then additional forms may also be possible.

Within this context, the Light-Utilising Nitrogenase (LUN) hypothesis becomes more plausible.

Alternative nitrogenases demonstrate that nitrogen fixation chemistry can be implemented through multiple catalytic architectures.

This evolutionary flexibility provides a foundation for future innovation.

7.14 Alternative Nitrogenases Beyond Earth

The significance of alternative nitrogenases may extend beyond Earth.

Different planetary environments possess different geochemistries.

Metal availability on other worlds may differ substantially from Earth.

Consequently:

Extraterrestrial life, if it exists, might evolve nitrogen-fixing systems unlike those known today.

Alternative nitrogenases therefore provide valuable models for astrobiology.

They illustrate how fundamental biological functions can be achieved through diverse chemical solutions.

7.15 Conclusions

Alternative nitrogenases reveal the remarkable adaptability of biological nitrogen fixation.

Key conclusions include:

1. Multiple nitrogenase systems exist in nature.
2. Vanadium and iron-only nitrogenases provide alternatives to molybdenum nitrogenase.
3. Alternative nitrogenases exhibit broader catalytic capabilities.
4. These systems preserve clues about ancient Earth environments.
5. Nitrogenase evolution reflects changing planetary geochemistry.
6. Alternative nitrogenases support the possibility of future catalytic innovations.
7. They provide important foundations for synthetic biology and the LUN hypothesis.

Within the Earth Clock framework, alternative nitrogenases are more than biochemical curiosities. They are evolutionary experiments that reveal how life repeatedly adapts to planetary change.

They remind us that evolution rarely follows a single path.

Instead, it explores many possibilities—some successful, some temporary, and some perhaps still waiting to be discovered.

End of Chapter 7

Chapter 8

Nitrogenase-Like Enzymes

Evolutionary Connections Between Nitrogen Fixation and Photosynthesis

8.1 Introduction

For many years, nitrogenase was regarded as a unique biological catalyst.

Its ability to reduce atmospheric nitrogen appeared unparalleled.

However, advances in comparative genomics, structural biology, and molecular evolution have revealed a surprising discovery:

Nitrogenase is not alone.

A growing family of nitrogenase-like enzymes has been identified throughout the biosphere.

These enzymes participate in:

- chlorophyll biosynthesis,
- bacteriochlorophyll biosynthesis,
- cofactor formation,
- tetrapyrrole metabolism,

- specialized redox reactions.

Although their biological functions differ, they share remarkable structural similarities with nitrogenase. These observations suggest that nitrogenase belongs to a much larger evolutionary superfamily.

Within the Earth Clock framework, these enzymes provide a bridge connecting two of Earth's most important biological innovations:

Nitrogen Fixation

and

Photosynthesis

8.2 The Nitrogenase Superfamily

The concept of a nitrogenase superfamily emerged when researchers discovered proteins exhibiting strong structural and sequence similarities to nitrogenase components.

Common characteristics include:

ATP-dependent electron transfer

Metal–sulfur clusters

Homologous protein architecture

Conserved catalytic mechanisms

Ancient evolutionary origins

These similarities indicate a shared ancestry.

Rather than evolving independently, many of these systems likely originated from a common ancestral metalloprotein.

8.3 Evolution Through Functional Diversification

Evolution frequently generates innovation through modification of existing systems.

A protein originally performing one function may gradually acquire a new role.

This process is known as:

Functional Diversification

Nitrogenase-like enzymes provide an excellent example.

A common ancestral system may have diversified into multiple catalytic lineages:

```
Ancient Metalloprotein
             |
     ---------------------
     |           |           |
Nitrogenase DPOR       COR
                          |
                        LPOR
```

Although simplified, this model illustrates how evolutionary diversification can generate entirely new biochemical functions.

8.4 DPOR:

Dark-Operative Protochlorophyllide Oxidoreductase

One of the most important nitrogenase-like enzymes is:

DPOR

(Dark-Operative Protochlorophyllide Oxidoreductase)

DPOR participates in chlorophyll biosynthesis.

Its function is to reduce protochlorophyllide during pigment formation.

Remarkably, DPOR shares extensive similarities with nitrogenase.

Structural Similarities

DPOR contains:

- ATP-dependent electron transfer proteins,
- iron–sulfur clusters,
- homologous catalytic architecture.

These features strongly suggest evolutionary relatedness.

Functional Differences

Unlike nitrogenase:

DPOR does not reduce nitrogen.

Instead, it catalyzes:

chlorophyll precursor reduction.

This represents one of the clearest examples of evolutionary functional divergence.

8.5 Why DPOR Matters

DPOR provides an important clue regarding the relationship between nitrogen fixation and photosynthesis.

Photosynthesis depends upon chlorophyll.

Chlorophyll synthesis depends partly upon DPOR.

DPOR appears evolutionarily related to nitrogenase.

This connection suggests that:

> The histories of nitrogen fixation and photosynthesis may be more closely linked than previously recognized.

Within the Earth Clock framework, this relationship is highly significant.

8.6 COR:

Chlorophyllide Oxidoreductase

Another important nitrogenase-like enzyme is:

COR

(Chlorophyllide Oxidoreductase)

COR participates in bacteriochlorophyll biosynthesis.

Like DPOR, COR shares:

- ATP-dependent electron transfer,
- iron–sulfur chemistry,
- structural homology.

COR further expands the nitrogenase superfamily.

Its existence demonstrates that nitrogenase-like systems have diversified extensively throughout evolutionary history.

8.7 Nitrogen Fixation and Photosynthesis: A Deep Evolutionary Connection

Nitrogen fixation and photosynthesis are often studied separately.

However, nitrogenase-like enzymes suggest a deeper relationship.

Consider the sequence:

Nitrogenase

Supports nitrogen acquisition.

↓

DPOR

Supports chlorophyll synthesis.

↓

Photosynthesis

Supports energy acquisition.

This progression implies that evolution repeatedly reused similar catalytic architectures to solve different biological challenges.

The result was two of the most transformative processes in Earth's history.

8.8 LPOR:

Light-Operative Protochlorophyllide Oxidoreductase

Perhaps the most intriguing member of this evolutionary story is:

LPOR

(Light-Operative Protochlorophyllide Oxidoreductase)

LPOR differs fundamentally from DPOR.

Unlike DPOR:

LPOR utilizes light energy directly.

Its reaction depends upon:

- photons,
- protochlorophyllide,
- NADPH.

Importantly:

LPOR does not require ATP-driven electron transfer machinery resembling nitrogenase.

This distinction has profound evolutionary implications.

8.9 The DPOR-to-LPOR Transition

The transition from DPOR-dominated systems to LPOR-dominated systems represents one of the most fascinating evolutionary transformations in biology.

Ancient photosynthetic organisms relied heavily upon DPOR.

Modern plants rely primarily upon LPOR.

This transition reflects a shift:

From:

ATP-dependent catalysis

To:

Light-dependent catalysis

The Earth Clock Theory views this transition as a major evolutionary event.

It demonstrates that evolution can replace ancient metabolic solutions with more efficient alternatives.

8.10 A Model for Evolutionary Innovation

The DPOR–LPOR transition provides an important conceptual model.

Evolutionary sequence:

Stage 1

ATP-dependent ancestral system.

↓

Stage 2

Intermediate diversification.

↓

Stage 3

Light-dependent system.

This pattern suggests that biological innovation often proceeds through gradual replacement rather than abrupt invention.

The same principle may apply to future nitrogen fixation systems.

8.11 The Evolutionary Basis of the LUN Hypothesis

The Light-Utilising Nitrogenase (LUN) hypothesis is built upon this evolutionary logic.

The argument is not that LUN already exists.

Rather:

The existence of LPOR demonstrates that evolution has previously transformed an ATP-dependent catalytic system into a light-dependent catalytic system.

Therefore:

A similar transformation might be possible for nitrogen fixation.

Conceptually:

Nitrogenase

↓

Nitrogenase-like Intermediate

↓

Light-Utilising Nitrogenase (LUN)

This possibility forms the evolutionary foundation of the LUN hypothesis.

8.12 Nitrogenase-Like Enzymes and Earth History

Nitrogenase-like enzymes provide a remarkable record of Earth's biological history.

Together they reveal:

- ancient metabolic innovation,
- chlorophyll evolution,
- photosynthetic diversification,
- adaptive responses to oxygenation.

Their evolutionary trajectories mirror major transitions in planetary history.

They are therefore valuable not only for molecular biology but also for understanding Earth's evolution as a whole.

8.13 Future Discoveries

Current knowledge likely represents only a fraction of existing diversity.

Future research may reveal:

New Nitrogenase-Like Enzymes

Novel Catalytic Functions

Additional Photosynthetic Connections

Previously Unknown Metal Cofactors

Transitional Evolutionary Forms

Such discoveries may further illuminate the relationship between nitrogen fixation and photosynthesis.

8.14 The Earth Clock Perspective

Within the Earth Clock framework, nitrogenase-like enzymes occupy a special position.

They connect:

- nutrient acquisition,
- energy acquisition,
- biological innovation.

Few enzyme families have influenced Earth so profoundly.

Nitrogenase enabled biological expansion.

Photosynthesis enabled planetary transformation.

Nitrogenase-like enzymes reveal the evolutionary bridge linking these processes.

Conclusions

Nitrogenase-like enzymes provide crucial insights into the evolution of life.

Key conclusions include:

1. Nitrogenase belongs to a larger enzyme superfamily.

2. DPOR and COR share deep evolutionary relationships with nitrogenase.
3. LPOR represents a fundamentally different, light-dependent catalytic strategy.
4. The DPOR-to-LPOR transition illustrates evolutionary replacement of ATP-dependent systems.
5. Nitrogen fixation and photosynthesis may share deeper evolutionary connections than previously recognized.
6. Nitrogenase-like enzymes provide a conceptual foundation for the LUN hypothesis.
7. These systems reveal how molecular innovation helped shape Earth's history.

Within the Earth Clock framework, nitrogenase-like enzymes represent one of the strongest examples of evolutionary continuity.

They demonstrate that nature rarely invents entirely new solutions.

Instead, it repeatedly modifies and repurposes successful molecular architectures.

The same principle may guide the future evolution of nitrogen fixation itself.

End of Chapter 8

Chapter 9

Photosynthesis and Planetary Oxygenation

The Second Great Biological Innovation

9.1 Introduction

If nitrogen fixation enabled biological expansion, photosynthesis enabled biological transformation.

Few biological processes have altered Earth as profoundly as photosynthesis.

Before photosynthesis:

- oxygen was scarce,
- ecosystems were limited,
- biological productivity remained relatively low.

After photosynthesis:

- atmospheric oxygen accumulated,
- food webs expanded,
- aerobic metabolism emerged,
- complex life became possible.

Within the Earth Clock framework, photosynthesis represents:

The Second Great Biological Innovation.

Nitrogen fixation solved the nitrogen problem.

Photosynthesis solved the energy problem.

Together they established the foundations of the modern biosphere.

9.2 The Energy Challenge

Early life existed in an energy-limited world.

Primitive organisms relied primarily upon:

- geochemical reactions,
- hydrothermal systems,
- fermentation,
- simple redox chemistry.

Although these processes supported microbial ecosystems, they restricted biological expansion.

The biosphere needed a larger and more reliable energy source.

That source was already available:

the Sun.

The challenge was learning how to use it.

9.3 The Origin of Photosynthesis

The earliest photosynthetic systems were likely simple.

They harvested light energy but did not produce oxygen.

These primitive organisms used electron donors such as:

- hydrogen sulfide,
- ferrous iron,
- organic compounds.

Generalized reaction:

$$CO2+2H2S+Light \rightarrow (CH2O)+H2O+2S$$

This process represented a major evolutionary advance.

For the first time, organisms could directly exploit solar energy.

However, a much greater innovation was still to come.

9.4 Cyanobacteria:

Architects of the Modern Atmosphere

The appearance of cyanobacteria transformed planetary history.

Unlike earlier photosynthetic organisms, cyanobacteria evolved the ability to use water as an electron donor.

Simplified reaction:

$$6CO_2+6H_2O+Light \rightarrow C_6H_{12}O_6+6O_2$$

This innovation had enormous consequences.

Water is abundant.

Using water removed a major ecological limitation.

More importantly:

oxygen became a by-product.

What initially appeared to be metabolic waste would eventually reshape the entire planet.

9.5 The Great Oxidation Event

One of the most important transitions in Earth history occurred approximately 2.4 billion years ago.

Scientists refer to this event as:

The Great Oxidation Event (GOE)

Before the GOE:

- atmospheric oxygen was extremely low,
- oceans contained abundant dissolved iron,
- anaerobic organisms dominated.

After the GOE:

- oxygen accumulated,
- atmospheric chemistry changed,
- ecosystems reorganized.

The Earth Clock Theory regards the GOE as a planetary turning point.

It marks the beginning of Earth's oxygen age.

9.6 Oxygen:

A Biological Revolution

Oxygen fundamentally altered life.

Advantages included:

High-Energy Metabolism

Aerobic respiration yields far more energy than anaerobic metabolism.

Larger Genomes

Greater energy availability supports more genetic complexity.

Multicellularity

Complex organisms become feasible.

Ecological Diversification

New environments become habitable.

These changes accelerated biological evolution.

The rise of oxygen transformed Earth from a microbial world into a platform for complexity.

9.7 Oxygen:

A Biological Crisis

Yet oxygen was not universally beneficial.

For many ancient organisms, oxygen was toxic.

The accumulation of oxygen likely caused:

- extinctions,
- ecological disruption,

- metabolic stress.

Some researchers describe the GOE as Earth's first environmental catastrophe.

Within the Earth Clock framework:

oxygen represents both opportunity and crisis.

Its emergence illustrates how evolutionary innovation often produces unintended consequences.

9.8 The Nitrogenase Paradox

Perhaps the most important consequence of oxygenation concerns nitrogen fixation.

Nitrogenase evolved in an anaerobic world.

Oxygen damages:

- iron–sulfur clusters,
- catalytic centers,
- electron transfer systems.

Thus the rise of oxygen created a paradox.

The process that enabled complex life simultaneously threatened one of the enzymes essential for sustaining life.

This challenge became one of the defining problems of biological evolution.

9.9 How Life Solved the Oxygen Problem

Evolution responded creatively.

Numerous strategies emerged.

Heterocysts

Specialized nitrogen-fixing cells in cyanobacteria.

Root Nodules

Symbiotic structures in legumes.

Temporal Separation

Photosynthesis during the day.

Nitrogen fixation at night.

Respiratory Protection

Rapid oxygen consumption near nitrogenase.

These innovations allowed nitrogen fixation and photosynthesis to coexist.

Together they sustained the expanding biosphere.

9.10 Photosynthesis and Nitrogen Fixation: The Dual Engine Hypothesis

The Earth Clock Theory proposes that modern life depends upon two foundational biochemical engines.

Photosynthesis

Provides:

- energy,
- biomass,
- oxygen.

Nitrogen Fixation

Provides:

- amino acids,
- proteins,
- nucleic acids.

Together they support nearly all biological productivity.

Neither system alone is sufficient.

Photosynthesis without nitrogen fixation eventually encounters nutrient limitation.

Nitrogen fixation without photosynthesis lacks adequate energy.

Their interaction shaped the entire course of biological evolution.

9.11 Planetary Consequences

The emergence of photosynthesis altered every major Earth system.

Atmosphere

Oxygen accumulated.

Oceans

Iron chemistry changed.

Climate

Greenhouse gas balance shifted.

Geology

Mineral diversity increased.

Biology

Complex ecosystems emerged.

Few biological processes have exerted such comprehensive influence.

Photosynthesis transformed Earth into a fundamentally different planet.

9.12 The Rise of Eukaryotes

One of the most important consequences of oxygenation was the emergence of eukaryotic life.

Higher energy availability supported:

- larger genomes,
- intracellular compartmentalization,
- increased cellular complexity.

Eukaryotes eventually gave rise to:

- algae,
- fungi,
- plants,
- animals.

The pathway toward complex life became increasingly accessible.

9.13 The Long Road to Complexity

Photosynthesis did not immediately produce animals or forests.

Instead, it initiated a long evolutionary process.

Over hundreds of millions of years:

- oxygen increased,
- ecosystems diversified,
- complexity accumulated.

The Earth Clock highlights the gradual nature of this transformation.

Major biological innovations often require immense spans of time.

Planetary evolution proceeds slowly but persistently.

9.14 Photosynthesis Beyond Earth

Photosynthesis may not be unique to Earth.

Many planets orbit stars capable of providing usable light energy.

If life exists elsewhere, photosynthesis or analogous systems may evolve repeatedly.

Such systems would offer major advantages:

- abundant energy,
- ecological expansion,
- long-term sustainability.

The Earth Clock Theory therefore predicts that photosynthesis-like processes may be common among inhabited worlds.

9.15 Toward the Green Planet

The success of photosynthesis eventually enabled the colonization of land.

Plants transformed terrestrial environments.

Forests emerged.

Soils developed.

Carbon cycling intensified.

A new phase of planetary evolution began.

The green planet we know today traces its origins to photosynthetic innovation billions of years ago.

Conclusions

Photosynthesis represents one of the most transformative innovations in Earth's history.

Key conclusions include:

1. Photosynthesis provided access to solar energy.
2. Cyanobacteria initiated planetary oxygenation.
3. The Great Oxidation Event transformed Earth's atmosphere.
4. Oxygen enabled high-energy metabolism and biological complexity.
5. Oxygen simultaneously challenged nitrogen fixation.
6. Evolution developed mechanisms allowing photosynthesis and nitrogen fixation to coexist.
7. Photosynthesis and nitrogen fixation function as the dual engines of the biosphere.

Within the Earth Clock framework, photosynthesis represents the Second Great Biological Innovation.

Nitrogen fixation enabled biological expansion.

Photosynthesis enabled planetary transformation.

Together they set the stage for the evolution of complex life.

Nitrogen built the biosphere.

Photosynthesis powered it.

End of Chapter 9

Chapter 10

DPOR, COR and LPOR

An Evolutionary Bridge Between Nitrogen Fixation and Photosynthesis

10.1 Introduction

Among all discoveries emerging from comparative molecular biology, few are more intriguing than the relationship between nitrogenase and chlorophyll biosynthesis enzymes.

For decades, nitrogen fixation and photosynthesis were regarded as independent biological processes.

Nitrogen fixation was viewed as a nutrient acquisition system.

Photosynthesis was viewed as an energy acquisition system.

However, structural and evolutionary analyses have revealed remarkable connections between these systems.

The key players are:

Nitrogenase

DPOR

(Dark-Operative Protochlorophyllide Oxidoreductase)

COR

(Chlorophyllide Oxidoreductase)

LPOR

(Light-Operative Protochlorophyllide Oxidoreductase)

Together these enzymes form an evolutionary narrative spanning billions of years.

Within the Earth Clock framework, they represent one of the clearest examples of molecular innovation shaping planetary evolution.

10.2 Chlorophyll:

The Pigment That Changed Earth

Photosynthesis depends upon chlorophyll.

Without chlorophyll:

- sunlight cannot be harvested efficiently,
- oxygenic photosynthesis cannot occur,

- modern ecosystems cannot exist.

Consequently, chlorophyll biosynthesis became one of the most important pathways in biological evolution.

The evolution of chlorophyll-producing systems transformed Earth from a microbial planet into a biologically complex world.

Understanding chlorophyll biosynthesis therefore helps explain how photosynthesis evolved.

10.3 DPOR:

The Nitrogenase-Like Chlorophyll Enzyme

DPOR catalyzes a critical step in chlorophyll biosynthesis.

Its function is:

Reduction of protochlorophyllide to chlorophyllide.

Remarkably, DPOR strongly resembles nitrogenase.

Structural Similarities

DPOR contains:

- ATP-dependent electron transfer proteins

- Iron–sulfur clusters
- Homologous catalytic architecture

Functional Similarities

Both systems:

- require ATP
- transfer electrons
- utilize metal clusters
- catalyze difficult reductions

These similarities strongly suggest common ancestry.

10.4 Molecular Evidence for Common Origins

Comparative studies reveal that DPOR proteins share significant similarities with nitrogenase proteins.

Examples include:

Nitrogenase	DPOR
NifH	BchL/ChlL
NifD	BchN/ChlN
NifK	BchB/ChlB

The parallels are striking.

Many researchers believe these systems evolved from an ancient common ancestor.

If correct, the connection between nitrogen fixation and photosynthesis extends deep into Earth's earliest history.

10.5 DPOR and the Ancient Earth

DPOR likely evolved in an anaerobic world.

Several observations support this conclusion.

Oxygen Sensitivity

DPOR is highly sensitive to oxygen.

Metal Cluster Dependence

Its catalytic machinery resembles ancient anaerobic enzymes.

Distribution

DPOR occurs primarily among:

- cyanobacteria
- algae

- photosynthetic bacteria

These characteristics suggest that DPOR emerged before widespread atmospheric oxygenation.

Within the Earth Clock framework, DPOR may represent a molecular relic of Earth's pre-oxygen world.

10.6 COR:

Expanding the Photosynthetic Toolkit

Another nitrogenase-like enzyme is COR.

COR participates in bacteriochlorophyll biosynthesis.

Functionally:

DPOR produces chlorophyll.

COR produces bacteriochlorophyll.

Although their products differ, both systems share:

- ATP dependence
- Iron–sulfur chemistry
- Nitrogenase-like architecture

COR therefore expands the nitrogenase superfamily.

Its existence demonstrates how evolution repeatedly reused successful catalytic frameworks.

10.7 Evolution Through Reuse

One of evolution's most important strategies is molecular reuse.

Rather than inventing entirely new proteins, evolution often modifies existing systems.

The process resembles engineering.

Successful designs become templates.

Nitrogenase-like enzymes provide an outstanding example.

A single ancestral catalytic architecture appears to have diversified into:

- nitrogen fixation
- chlorophyll biosynthesis
- bacteriochlorophyll biosynthesis

The result was a cascade of innovations that transformed Earth's biosphere.

10.8 LPOR:

A Revolutionary New Strategy

The emergence of LPOR represents a major departure from previous systems.

Unlike DPOR and COR:

LPOR directly utilizes light energy.

Its reaction depends upon:

- photons
- protochlorophyllide
- NADPH

Importantly:

LPOR does not require ATP-dependent electron transfer machinery resembling nitrogenase.

This distinction fundamentally changed chlorophyll biosynthesis.

10.9 From ATP to Light

The transition from DPOR to LPOR represents a remarkable evolutionary event.

DPOR strategy:

ATP-driven reduction.

LPOR strategy:

Light-driven reduction.

Conceptually:

DPOR

ATP → Electron Transfer → Reduction

↓

LPOR

Light → Reduction

This transformation illustrates how evolution can replace ancient metabolic solutions with more efficient alternatives.

10.10 Why LPOR Won

LPOR possesses several advantages.

Oxygen Tolerance

Unlike DPOR, LPOR functions effectively in oxygen-rich environments.

Energy Efficiency

Light directly powers catalysis.

Simplicity

LPOR requires fewer accessory components.

Compatibility

LPOR integrates naturally into modern photosynthetic systems.

These advantages explain why LPOR dominates in higher plants today.

10.11 The Evolutionary Sequence

The Earth Clock Theory proposes a simplified evolutionary progression:

Ancient Metalloprotein

↓

Nitrogenase

↓

DPOR

↓

COR

↓

LPOR

Each stage represents a major innovation.

Nitrogenase

Nitrogen acquisition.

DPOR

Primitive chlorophyll biosynthesis.

COR

Diversified photosynthetic pigments.

LPOR

Direct photochemical catalysis.

This sequence illustrates increasing integration between energy acquisition and biological function.

10.12 The Birth of a New Concept

The DPOR-to-LPOR transition suggests a broader principle.

Evolution tends toward:

- improved efficiency,
- reduced complexity,

- greater environmental compatibility.

If ATP-dependent chlorophyll biosynthesis evolved into light-dependent chlorophyll biosynthesis, could ATP-dependent nitrogen fixation eventually evolve into light-dependent nitrogen fixation?

This question forms the basis of the LUN hypothesis.

10.13 The LUN Evolutionary Framework

The Earth Clock Theory proposes a conceptual pathway:

Nitrogenase

↓

Nitrogenase-like Intermediate

↓

Photochemical Nitrogenase

↓

LUN

No such system has yet been discovered.

However, LPOR demonstrates that analogous transitions are biologically possible.

Evolution has already performed a similar transformation once.

10.14 Photosynthesis and Nitrogen Fixation: Converging Histories

Traditionally, photosynthesis and nitrogen fixation have been studied independently.

Yet nitrogenase-like enzymes reveal profound connections.

Photosynthesis and nitrogen fixation share:

- evolutionary ancestry
- catalytic principles
- metal-cluster chemistry
- planetary significance

The two systems may represent branches of a common evolutionary story.

Within the Earth Clock framework:

they become parallel solutions to two fundamental challenges.

Photosynthesis

How can life acquire energy?

Nitrogen Fixation

How can life acquire nitrogen?

Together these solutions enabled Earth's biological expansion.

10.15 The Earth Clock Perspective

The Earth Clock Theory interprets DPOR, COR and LPOR as milestones in molecular evolution.

Each innovation expanded biological possibilities.

Nitrogenase enabled atmospheric nitrogen utilization.

DPOR enabled chlorophyll synthesis.

COR diversified photosynthetic strategies.

LPOR linked chlorophyll production directly to light.

This progression reflects a recurring theme in evolution:

> increasing integration between energy, information and metabolism.

10.16 Implications for Future Biology

Understanding these evolutionary transitions has practical implications.

Future research may enable:

Artificial Chloroplast Engineering

Synthetic Nitroplasts

Light-Driven Nitrogen Fixation

AI-Guided Enzyme Design

New Catalytic Architectures

The lessons of ancient evolution may guide future biotechnology.

Conclusions

DPOR, COR and LPOR provide critical insights into the relationship between nitrogen fixation and photosynthesis.

Key conclusions include:

1. DPOR and COR belong to the nitrogenase superfamily.
2. These enzymes preserve evidence of shared evolutionary origins.

3. LPOR represents a fundamentally different light-driven strategy.
4. The DPOR-to-LPOR transition demonstrates evolutionary replacement of ATP-dependent systems.
5. Photosynthesis and nitrogen fixation may share deeper evolutionary connections than previously recognized.
6. These transitions provide the conceptual foundation for the LUN hypothesis.
7. Molecular evolution helped shape planetary evolution.

Within the Earth Clock framework, DPOR, COR and LPOR represent an evolutionary bridge connecting the two greatest biochemical innovations in Earth's history:

Nitrogen Fixation and **Photosynthesis**.

Nitrogenase helped life acquire nitrogen.

LPOR helped life acquire light.

Together they helped transform Earth.

End of Chapter 10

Chapter 11

Oxygen: Friend and Enemy

The Oxygen Paradox in the Evolution of Life

11.1 Introduction

Few substances have influenced the history of Earth more profoundly than oxygen.

Today oxygen is essential for:

- humans,
- animals,
- plants,
- fungi,
- countless microorganisms.

Every breath taken by an animal reflects a biological dependency that evolved over billions of years.

Yet oxygen was not always beneficial.

For much of Earth's history, oxygen was either absent or extremely rare.

When oxygen first accumulated in significant quantities, it triggered one of the greatest environmental crises ever experienced by life.

Within the Earth Clock framework, oxygen represents a remarkable paradox:

> The molecule that enabled complex life was also one of the greatest threats to ancient life.

This contradiction lies at the heart of biological evolution.

11.2 The Oxygen-Free Earth

During the first half of Earth's history, atmospheric oxygen was nearly absent.

The ancient atmosphere consisted largely of:

- nitrogen,
- carbon dioxide,
- methane,
- water vapor.

Under these conditions:

- anaerobic metabolism dominated,
- iron remained soluble,
- oxygen-sensitive enzymes flourished.

Nitrogenase evolved within this world.

The enzyme's extreme oxygen sensitivity suggests an origin long before atmospheric oxygen became abundant.

In many ways, nitrogenase remains a molecular relic of the pre-oxygen Earth.

11.3 The Rise of Oxygen

The evolution of oxygenic photosynthesis fundamentally altered Earth's atmosphere.

Cyanobacteria began using water as an electron donor:

6CO_2+6H_2O+Light\rightarrow

C_6H_{12}O_6+6O_2

Initially, most oxygen reacted with:

- dissolved iron,
- volcanic gases,
- reduced minerals.

For millions of years, oxygen accumulation remained limited.

Eventually, however, geological sinks became saturated.

Oxygen began accumulating in the atmosphere.

Earth entered a new evolutionary era.

11.4 The Great Oxidation Event

Approximately 2.4 billion years ago, oxygen levels increased dramatically.

This transition is known as:

The Great Oxidation Event (GOE)

The GOE altered:

Atmospheric Chemistry

Ocean Chemistry

Climate

Biological Evolution

Few events have had such far-reaching consequences.

Within the Earth Clock framework, the GOE represents one of the most important turning points in planetary history.

11.5 The Oxygen Catastrophe

For many ancient organisms, oxygen was toxic.

Reactive oxygen species can damage:

- DNA,
- proteins,
- membranes,
- metal cofactors.

Consequently, oxygen accumulation likely caused widespread ecological disruption.

Many anaerobic organisms:

- retreated to protected environments,
- adapted,
- or became extinct.

This event is sometimes described as:

The Oxygen Catastrophe

A planetary environmental crisis created entirely by biological activity.

It represents one of the earliest examples of life reshaping its own environment.

11.6 Oxygen and Energy

Although oxygen posed challenges, it also created extraordinary opportunities.

Aerobic respiration generates substantially more energy than anaerobic metabolism.

Simplified reaction:

$C_6H_{12}O_6+6O_2\rightarrow6CO_2+6H_2O+$ Energy

Compared with fermentation:

Aerobic respiration yields far greater ATP production.

Consequences included:

- larger cells,
- larger genomes,
- greater complexity,
- multicellularity.

Oxygen therefore became a major driver of evolutionary innovation.

11.7 The Nitrogenase Problem

The rise of oxygen created a serious challenge for nitrogen fixation.

Nitrogenase contains oxygen-sensitive metal clusters.

Exposure to oxygen can:

- oxidize catalytic centers,
- disrupt electron transfer,
- permanently inactivate enzymes.

Thus oxygen and nitrogenase became natural enemies.

This conflict created one of evolution's most persistent challenges.

How can organisms perform nitrogen fixation in an oxygen-rich world?

11.8 The Oxygen Paradox Theory

The Earth Clock Theory introduces:

The Oxygen Paradox Theory

The theory states:

> Oxygen simultaneously promotes and restricts biological complexity.

Oxygen Promotes Complexity

Through:

- aerobic respiration,
- energy production,
- ecological expansion.

Oxygen Restricts Complexity

Through:

- oxidative stress,
- enzyme damage,
- nitrogenase inhibition.

Thus oxygen acts as both:

Friend

and

Enemy

of life.

The entire history of complex life reflects attempts to balance these opposing effects.

11.9 Evolutionary Solutions

Evolution developed numerous strategies to protect oxygen-sensitive processes.

Among the most important are mechanisms protecting nitrogen fixation.

These solutions represent some of the most elegant innovations in biology.

11.10 Heterocysts

Certain cyanobacteria evolved specialized cells called:

Heterocysts

Functions:

- exclude oxygen,
- maintain anaerobic conditions,
- support nitrogen fixation.

Heterocysts separate:

Photosynthesis

from

Nitrogen Fixation

within multicellular microbial filaments.

This represents one of nature's earliest examples of cellular division of labor.

11.11 Temporal Separation

Some organisms solve the oxygen problem through timing.

Strategy:

Daytime

Photosynthesis

Nighttime

Nitrogen Fixation

This temporal separation minimizes oxygen exposure.

The approach demonstrates that evolution can use time as a protective mechanism.

11.12 Root Nodules

Legume–rhizobium symbioses represent another solution.

Root nodules create microenvironments favorable for nitrogen fixation.

Key feature:

Leghemoglobin

Leghemoglobin regulates oxygen availability.

It supplies sufficient oxygen for respiration while protecting nitrogenase.

This elegant system illustrates how evolution can reconcile apparently incompatible processes.

11.13 Respiratory Protection

Some free-living nitrogen-fixing bacteria utilize intense respiration to consume oxygen rapidly.

This strategy creates localized low-oxygen environments.

Although energetically expensive, it permits nitrogen fixation under otherwise unfavorable conditions.

The approach demonstrates the importance of nitrogen fixation.

Evolution repeatedly invests substantial resources to preserve this capability.

11.14 Oxygen and Complex Life

The rise of oxygen enabled:

- eukaryotes,
- multicellularity,
- animals,

- terrestrial ecosystems.

Without oxygen:

complex life as we know it would almost certainly not exist.

Yet oxygen alone was insufficient.

Complex life also required:

- nitrogen,
- photosynthesis,
- ecological stability.

Thus oxygen became one component of a larger evolutionary system.

11.15 The Long Coexistence

For over two billion years:

oxygen production and nitrogen fixation have coexisted.

This coexistence was never inevitable.

It required continuous evolutionary innovation.

The Earth Clock framework emphasizes that much of biological history can be interpreted as a long-term negotiation between:

Oxygen

and

Nitrogenase

The success of modern ecosystems reflects the outcome of that negotiation.

11.16 Oxygen Beyond Earth

The Oxygen Paradox may apply beyond Earth.

Any planet supporting oxygen-producing organisms may eventually encounter similar challenges.

Future extraterrestrial biospheres could face:

- oxidative stress,
- metabolic conflicts,
- nutrient limitations.

The Earth Clock Theory therefore suggests that oxygen-related evolutionary pressures may be universal among complex biospheres.

11.17 Oxygen and the Future

Modern civilization is increasingly capable of modifying atmospheric composition.

Future technologies may influence:

- carbon cycling,
- oxygen production,
- ecosystem function.

Understanding oxygen's dual role becomes increasingly important.

The same molecule that once transformed Earth continues to influence its future.

Conclusions

Oxygen represents one of the most influential molecules in Earth's history.

Key conclusions include:

1. Oxygen was initially absent from Earth's atmosphere.
2. Oxygenic photosynthesis transformed planetary chemistry.
3. The Great Oxidation Event altered biological evolution.
4. Oxygen enabled high-energy metabolism and complexity.

5. Oxygen simultaneously threatened nitrogen fixation.
6. Evolution developed numerous protective mechanisms.
7. The Oxygen Paradox continues to shape life today.

Within the Earth Clock framework, oxygen is both creator and destroyer.

It enabled the rise of complex life.

Yet it continually challenged one of the most ancient and important enzymes on Earth.

Oxygen gave life greater power.

Nitrogenase ensured that life could continue to grow.

The history of life is the story of their coexistence.

End of Chapter 11

Chapter 12

Nitrogen Fixation and Complex Life

Nitrogen as a Central Constraint on Biological Evolution

12.1 Introduction

The history of life is often told as a story of increasing complexity.

From simple cells emerged:

- multicellular organisms,
- plants,
- animals,
- ecosystems,
- civilizations.

Most explanations emphasize:

- energy availability,
- genetic innovation,
- natural selection,
- environmental change.

While all are important, another factor may have been equally fundamental:

Nitrogen Availability

Every increase in biological complexity requires additional biological material.

Proteins increase.

Genomes expand.

Cells differentiate.

Tissues specialize.

All of these processes require nitrogen.

Within the Earth Clock framework:

> Nitrogen availability acted as a major constraint on the evolution of complex life.

12.2 Complexity Requires Matter

Biologists often focus on information.

Genes encode biological structures.

Natural selection shapes adaptation.

However, information alone is insufficient.

Every gene requires physical molecules.

Every molecule requires atoms.

Among those atoms, nitrogen occupies a special position.

Nitrogen is essential for:

- amino acids,
- nucleotides,
- ATP,
- chlorophyll,
- enzymes.

Without nitrogen:

information cannot become biology.

12.3 The Nitrogen Threshold Hypothesis

The Earth Clock Theory introduces:

The Nitrogen Threshold Hypothesis

The hypothesis states:

> Major increases in biological complexity become possible only after sufficient fixed nitrogen becomes available.

Under this model:

Low Nitrogen Availability

Supports:

- small populations,
- limited genomes,

- simple ecosystems.

High Nitrogen Availability

Supports:

- larger populations,
- greater biodiversity,
- more complex organisms.

Nitrogen therefore functions as a biological threshold variable.

Crossing certain nitrogen thresholds may enable entirely new evolutionary possibilities.

12.4 Nitrogen and Genome Expansion

One of the most important transitions in evolution was the expansion of genomes.

Early organisms likely possessed relatively small genomes.

As evolution progressed:

- new genes appeared,
- regulatory networks expanded,
- developmental complexity increased.

Every nucleotide contains nitrogen.

Consequently:

larger genomes require greater nitrogen investment.

This relationship is rarely emphasized.

Yet it suggests a direct connection between nitrogen availability and biological information capacity.

Within the Earth Clock framework:

> Nitrogen helped determine how much biological information Earth could support.

12.5 Nitrogen and Protein Diversity

Complex organisms require enormous numbers of proteins.

Examples include:

- enzymes,
- receptors,
- transporters,
- structural proteins,
- signaling molecules.

Proteins are constructed from amino acids.

Every amino acid contains nitrogen.

As protein diversity increases:

nitrogen demand increases.

The evolution of sophisticated metabolism therefore depends partly upon the availability of fixed nitrogen.

12.6 Nitrogen and Multicellularity

The emergence of multicellular organisms represents one of the greatest transitions in evolutionary history.

Multicellularity requires:

- cell differentiation,
- communication,
- structural support,
- developmental regulation.

All of these processes increase nitrogen requirements.

A multicellular organism contains vastly more nitrogen-rich biomolecules than a single cell.

The Earth Clock Theory therefore proposes:

> The evolution of multicellularity depended not only upon energy availability but also upon sufficient nitrogen availability.

12.7 Nitrogen and Plant Evolution

Plants transformed Earth's surface.

The colonization of land required:

- chlorophyll,
- photosynthetic machinery,
- structural tissues,
- reproductive systems.

All depend heavily upon nitrogen.

Modern agriculture clearly demonstrates this relationship.

Plant growth often responds directly to nitrogen availability.

The same principle likely influenced evolutionary history.

The diversification of terrestrial plants may have been facilitated by expanding nitrogen supplies.

12.8 Nitrogen and Animal Evolution

Animals represent another major increase in biological complexity.

Compared with microbes:

animals require:

- nervous systems,
- muscles,
- immune systems,
- reproductive organs.

These structures contain enormous quantities of proteins.

Protein synthesis requires nitrogen.

Thus animal evolution indirectly depends upon nitrogen fixation occurring elsewhere in the biosphere.

Every animal ecosystem ultimately traces its nitrogen back to microbial activity.

12.9 Food Webs and Nitrogen Flow

Nitrogen moves through ecosystems in predictable ways.

General pathway:

Atmospheric Nitrogen

↓

Nitrogen Fixation

↓

Plants

↓

Herbivores

↓

Carnivores

↓

Decomposition

↓

Nitrogen Recycling

Nitrogen fixation forms the gateway.

Without it:

the entire food web collapses.

The importance of nitrogen fixation therefore extends far beyond microbes.

It supports every trophic level.

12.10 Nitrogen and Ecosystem Size

Nitrogen availability strongly influences ecosystem productivity.

Examples include:

Tropical Forests

High nitrogen cycling supports enormous biomass.

Grasslands

Nitrogen often limits productivity.

Oceans

Nitrogen frequently controls primary production.

These patterns suggest that nitrogen has regulated ecosystem size throughout Earth's history.
The same principle likely applied during ancient evolutionary transitions.

12.11 Nitrogen and Biodiversity

Biodiversity depends upon ecological opportunity.

Nitrogen availability influences:

- plant productivity,
- habitat complexity,
- food availability,
- population size.

Greater nitrogen availability often supports greater biodiversity.

Within the Earth Clock framework:

nitrogen fixation indirectly contributed to the diversification of life.

By increasing ecological carrying capacity, it created opportunities for evolutionary innovation.

12.12 Nitrogen as Biological Infrastructure

Modern civilization depends upon infrastructure.

Roads.

Power grids.

Communication networks.

Similarly, ecosystems depend upon biological infrastructure.

Nitrogen fixation represents one of the most important components of that infrastructure.

It continuously supplies the raw materials necessary for life.

Most organisms never fix nitrogen themselves.

Yet all depend upon organisms that do.

Nitrogen fixation therefore functions as a hidden foundation beneath the entire biosphere.

12.13 The Evolutionary Cascade

The Earth Clock Theory proposes an evolutionary cascade:

Nitrogen Fixation

↓

More Fixed Nitrogen

↓

Greater Biomass

↓

Larger Populations

↓

Greater Genetic Diversity

↓

Increased Complexity

↓

Complex Life

This sequence does not imply that nitrogen alone caused complexity.

Rather:

nitrogen availability created conditions under which complexity became possible.

12.14 Revisiting Earth's History

Viewed through this lens, many major evolutionary events acquire new meaning.

Emergence of Eukaryotes

Required increased metabolic and genetic complexity.

Multicellularity

Required expanded protein synthesis.

Land Plants

Required large-scale biomass production.

Animals

Required sophisticated physiological systems.

Each transition increased nitrogen demand.

This recurring pattern supports the Nitrogen Threshold Hypothesis.

12.15 The Central Constraint Concept

The Earth Clock Theory proposes:

> Nitrogen fixation acted as a central constraint on biological evolution.

A constraint is not necessarily a direct cause.

Rather:

it determines what becomes possible.

Examples:

Energy constrains metabolism.

Water constrains ecosystems.

Nitrogen constrains complexity.

This perspective places nitrogen fixation among the most influential processes in Earth's history.

12.16 Humanity and Nitrogen

Human civilization provides a modern example.

Agricultural productivity depends heavily upon nitrogen availability.

Population growth has closely tracked increases in fixed nitrogen supply.

The Haber–Bosch process dramatically expanded human carrying capacity.

This relationship demonstrates that nitrogen remains a central constraint even today.

The same principle likely operated throughout evolutionary history.

12.17 Nitrogen Beyond Earth

The significance of nitrogen may extend far beyond our planet.

Any biosphere based upon:

- proteins,
- nucleic acids,
- cellular metabolism,

will require nitrogen.

Consequently:

Nitrogen thresholds may represent universal features of biological evolution.

The Earth Clock Theory predicts that nitrogen acquisition systems could become critical milestones on many inhabited worlds.

12.18 Toward a Unified Theory

Photosynthesis solved the energy problem.

Nitrogen fixation solved the nitrogen problem.

Together they created conditions for complex life.

The Earth Clock framework therefore proposes a unified interpretation:

Energy Enables

Nitrogen Builds

Photosynthesis provides energy.

Nitrogen fixation provides biological construction material.

Complex life emerges from their interaction.

Conclusions

Nitrogen fixation played a central role in the evolution of complexity.

Key conclusions include:

1. Complex life requires large amounts of nitrogen.
2. Genome expansion increases nitrogen demand.
3. Protein diversity depends upon fixed nitrogen availability.
4. Multicellularity requires substantial nitrogen investment.
5. Ecosystem productivity is strongly influenced by nitrogen cycling.
6. Nitrogen fixation functions as foundational biological infrastructure.
7. Nitrogen availability may have acted as a central evolutionary constraint.

Within the Earth Clock framework, nitrogen fixation is not merely a microbial process.

It is one of the fundamental mechanisms that allowed Earth to evolve from a planet of microbes into a planet of forests, animals and civilizations.

Photosynthesis supplied the energy.

Nitrogen fixation supplied the materials.

Together they built the living Earth.

End of Chapter 12

Chapter 13

Engineering Nitrogen Fixation

One of the Grand Challenges of the Twenty-First Century

13.1 Introduction

For billions of years, biological nitrogen fixation evolved through natural selection.

Microorganisms developed nitrogenase.

Symbiotic systems emerged.

Complex ecological networks formed.

Yet most crop plants remain unable to fix atmospheric nitrogen independently.

Modern agriculture therefore depends heavily upon industrial nitrogen fertilizers.

This dependence creates enormous challenges:

- energy consumption,
- greenhouse gas emissions,
- water pollution,
- ecosystem disruption.

The ability to engineer biological nitrogen fixation into major crops has become one of the most ambitious goals in modern biology.

Within the Earth Clock framework, engineering nitrogen fixation represents a new phase of evolution:

> The transition from natural nitrogen fixation to intentional nitrogen fixation.

13.2 The Haber–Bosch Era

The twentieth century witnessed one of the greatest technological achievements in human history.

The Haber–Bosch process enabled industrial nitrogen fixation.

Simplified reaction:

$N_2+3H_2\rightarrow2NH_3$

This innovation dramatically increased food production.

Today, approximately half of the nitrogen present in human bodies can be traced to Haber–Bosch-derived fertilizer.

Yet this success carries significant costs.

Industrial nitrogen fixation requires:

- high temperatures,
- high pressures,
- fossil-fuel inputs.

The search for sustainable alternatives continues.

13.3 Why Engineering Nitrogen Fixation Matters

Global agriculture faces unprecedented challenges.

Population Growth

More food is required.

Climate Change

Agricultural systems must become more resilient.

Environmental Protection

Nitrogen pollution must be reduced.

Sustainability

Dependence on industrial fertilizers remains problematic.

Engineering biological nitrogen fixation offers a potential solution to all four challenges.

13.4 The Nitrogen-Fixing Cereals Dream

One of the oldest goals in agricultural biotechnology is:

Nitrogen-Fixing Cereals

Imagine:

- wheat fixing nitrogen,
- rice fixing nitrogen,
- maize fixing nitrogen.

Such crops could dramatically reduce fertilizer dependence.

The concept has inspired researchers for decades.

However, implementation remains difficult.

Nitrogen fixation is not controlled by a single gene.

It requires an entire molecular system.

13.5 The Complexity Problem

Nitrogenase is among the most complicated enzymes known.

Successful nitrogen fixation requires:

Structural Genes

nifHDK and associated genes.

Cofactor Assembly

Construction of FeMoco.

Electron Transfer Systems

Specialized reducing pathways.

Regulatory Networks

Precise control mechanisms.

Oxygen Protection

Maintenance of favorable environments.

The challenge is therefore far greater than introducing a single trait.

Engineering nitrogen fixation requires reconstruction of an entire biological process.

13.6 Symbiotic Engineering

One strategy focuses on improving existing symbioses.

Legumes already form nitrogen-fixing associations with rhizobia.

Researchers aim to extend similar interactions to non-legume crops.

Approaches include:

Enhanced Microbial Colonization

Synthetic Symbioses

Engineered Root Systems

Novel Signaling Pathways

This strategy works with natural biological systems rather than replacing them.

13.7 Endophytic Nitrogen Fixation

Certain microorganisms live inside plant tissues.

These organisms are known as:

Endophytes

Some possess nitrogen-fixing capability.

Advantages include:

- close association with crops,
- reduced oxygen exposure,
- direct nutrient transfer.

Engineering endophytic systems may provide practical pathways toward improved biological nitrogen fixation.

13.8 Chloroplast Engineering

One of the most exciting possibilities involves chloroplasts.

Chloroplasts possess several attractive features:

High Gene Expression

Energy Availability

Relative Genetic Independence

Evolutionary Origin from Bacteria

Because chloroplasts evolved from cyanobacteria, they retain characteristics favorable for engineering.

Many researchers view chloroplasts as promising platforms for future nitrogen fixation systems.

13.9 Why Chloroplasts Are Interesting

Chloroplasts already perform:

- photosynthesis,
- electron transfer,
- ATP generation.

These functions overlap conceptually with some requirements of nitrogen fixation.

Potential advantages include:

Direct Energy Supply

Integrated Metabolism

Efficient Resource Allocation

However, chloroplasts also contain oxygen.

This creates a major challenge.

13.10 The Oxygen Barrier

Oxygen remains the central obstacle.

Photosynthesis generates oxygen.

Nitrogenase is oxygen sensitive.

This contradiction represents one of the greatest engineering challenges in biology.

Potential solutions include:

Temporal Separation

Spatial Separation

Protective Compartments

Oxygen-Tolerant Variants

Synthetic Organelles

Future breakthroughs may depend upon solving this problem.

13.11 Synthetic Biology

Recent advances in synthetic biology have transformed biological engineering.

Scientists can now:

- design pathways,
- assemble genomes,
- construct synthetic circuits,
- redesign metabolic networks.

Nitrogen fixation has become a major target.

Synthetic biology enables researchers to move beyond natural evolutionary constraints.

For the first time, entirely new nitrogen-fixing systems may become possible.

13.12 Artificial Intelligence and Protein Design

Artificial intelligence has introduced powerful new capabilities.

Modern tools include:

- AlphaFold
- RoseTTAFold
- RFdiffusion
- ProteinMPNN
- LigandMPNN

These systems allow scientists to:

Predict Structures

Design New Proteins

Optimize Electron Transfer

Improve Stability

Explore Novel Catalytic Architectures

AI may dramatically accelerate efforts to engineer nitrogen fixation.

13.13 Synthetic Nitroplasts

A particularly ambitious concept is the creation of:

Synthetic Nitroplasts

Nitroplasts are hypothetical nitrogen-fixing organelles.

Like chloroplasts and mitochondria, they would function as specialized intracellular compartments.

Potential functions:

- nitrogen fixation,
- ammonia production,
- metabolic integration.

Although currently theoretical, nitroplasts represent one of the most exciting future directions.

13.14 Engineering Evolution

Traditional evolution operates through:

- mutation,
- selection,
- time.

Engineering introduces a new process:

Directed Evolution

Scientists can now accelerate adaptation.

Benefits include:

- faster optimization,

- targeted modification,
- exploration of new catalytic space.

The Earth Clock framework views this transition as highly significant.

Evolution itself is becoming an engineering process.

13.15 Agricultural Transformation

Successful nitrogen-fixing crops could revolutionize agriculture.

Potential outcomes include:

Reduced Fertilizer Use

Lower Production Costs

Improved Sustainability

Reduced Environmental Damage

Increased Food Security

The implications extend far beyond biology.

They affect economics, society and global development.

13.16 Challenges Ahead

Despite enormous progress, significant obstacles remain.

These include:

Oxygen Sensitivity

Genetic Complexity

Cofactor Assembly

Metabolic Integration

Ecological Stability

No single solution currently addresses all challenges.

Future success will likely require multiple complementary strategies.

13.17 The Earth Clock Perspective

Within the Earth Clock framework, engineering nitrogen fixation represents a new evolutionary stage.

Past:

Natural evolution created nitrogen fixation.

Present:

Humanity studies nitrogen fixation.

Future:

Humanity may redesign nitrogen fixation.

This progression reflects increasing biological self-awareness.

For the first time, life is attempting to consciously modify one of its most ancient metabolic systems.

13.18 Toward the Next Biological Revolution

The successful engineering of nitrogen fixation could become one of the defining scientific achievements of the twenty-first century.

Its significance may rival:

- agriculture,
- photosynthesis research,
- genetic engineering.

Such a breakthrough would not merely improve crops.

It would alter the relationship between humanity and the nitrogen cycle itself.

Conclusions

Engineering nitrogen fixation remains one of the grand challenges of modern science.

Key conclusions include:

1. Modern agriculture depends heavily upon industrial nitrogen fixation.
2. Sustainable alternatives are urgently needed.
3. Nitrogen-fixing cereals remain a major objective.
4. Chloroplast engineering offers promising opportunities.
5. Oxygen sensitivity remains the central obstacle.
6. Synthetic biology and AI are accelerating progress.
7. Synthetic nitroplasts may represent future solutions.

Within the Earth Clock framework, engineering nitrogen fixation marks the beginning of a new era.

For billions of years, nitrogen fixation evolved naturally.

Humanity may now become an active participant in its future evolution.

Evolution created nitrogen fixation.

Science seeks to improve it.

The future may redefine it.

End of Chapter 13

Chapter 14

Synthetic Nitroplasts

Engineering the Next Organelle of Life

14.1 Introduction

The history of life can be viewed as a history of cellular innovation.

Among the most important innovations were the emergence of biological organelles.

Two organelles fundamentally transformed the evolution of complex life:

Mitochondria

and

Chloroplasts

Mitochondria enabled efficient energy production.

Chloroplasts enabled photosynthesis.

Together they reshaped Earth's biosphere.

The Earth Clock Theory proposes that a third major organelle may eventually emerge:

The Nitroplast

A nitrogen-fixing organelle capable of providing bioavailable nitrogen directly within eukaryotic cells. Although still largely theoretical, the nitroplast concept represents one of the most exciting possibilities in future biology.

14.2 The Endosymbiotic Revolution

Modern biology recognizes that mitochondria and chloroplasts originated through endosymbiosis.

The basic concept is simple.

An ancestral cell engulfed another microorganism.

Instead of being digested, the engulfed organism survived.

Over evolutionary time:

- cooperation increased,
- gene transfer occurred,
- metabolic integration developed.

Eventually a new organelle emerged.

This process fundamentally altered life's trajectory.

14.3 The Origin of Mitochondria

Mitochondria evolved from ancient bacteria related to modern alphaproteobacteria.

Their emergence provided:

Efficient ATP Production

Enhanced Metabolism

Larger Cell Size

Increased Genetic Complexity

Without mitochondria, complex eukaryotic life would likely never have evolved.

The mitochondrial revolution transformed biological energy management.

14.4 The Origin of Chloroplasts

A second endosymbiotic event produced chloroplasts.

An ancestral eukaryotic cell incorporated a cyanobacterium.

The result was transformative.

Chloroplasts enabled:

- photosynthesis,

- solar energy capture,
- large-scale biomass production.

This innovation ultimately produced:

- algae,
- forests,
- terrestrial ecosystems.

The chloroplast revolution transformed planetary productivity.

14.5 A Pattern in Evolution

The emergence of mitochondria and chloroplasts reveals an important evolutionary pattern.

Major biological innovations often arise through:

Integration

rather than

Invention

Instead of creating entirely new systems, evolution combines existing systems in novel ways.

The Earth Clock Theory proposes that nitrogen fixation may eventually follow a similar path.

14.6 Why Nitrogen Fixation Is Different

Unlike photosynthesis and respiration, nitrogen fixation remains largely restricted to prokaryotes.

Most plants cannot directly fix atmospheric nitrogen.

Instead they depend upon:

- symbioses,
- microbial partners,
- soil nitrogen.

This dependence creates major limitations.

A nitrogen-fixing organelle could potentially remove these constraints.

14.7 The Nitroplast Concept

The nitroplast can be defined as:

> A specialized intracellular organelle dedicated to biological nitrogen fixation.

Conceptually, a nitroplast would function similarly to:

Mitochondria

Energy acquisition.

Chloroplasts

Light acquisition.

Nitroplasts

Nitrogen acquisition.

Such an organelle would fundamentally alter cellular metabolism.

14.8 Natural Precedents

Recent discoveries suggest that nitrogen-fixing endosymbioses may already exist in nature.

Certain marine microorganisms exhibit highly integrated relationships with nitrogen-fixing partners.

These systems provide valuable models.

They demonstrate that evolutionary integration between host cells and nitrogen-fixing organisms is biologically possible.

Such observations strengthen the conceptual foundation of nitroplast evolution.

14.9 The Pathway to a Nitroplast

A hypothetical evolutionary pathway may involve several stages.

Stage 1

Free-living diazotroph.

↓

Stage 2

Stable symbiosis.

↓

Stage 3

Intracellular residence.

↓

Stage 4

Genome reduction.

↓

Stage 5

Metabolic integration.

↓

Stage 6

True nitroplast.

This sequence resembles the evolutionary histories of mitochondria and chloroplasts.

14.10 Engineering a Synthetic Nitroplast

Natural evolution may require millions of years.

Synthetic biology could accelerate the process dramatically.

Potential strategies include:

Engineering Diazotrophic Endosymbionts

Genome Minimization

Controlled Gene Transfer

Metabolic Integration

Artificial Compartmentalization

These approaches aim to reproduce evolutionary processes within experimental timescales.

14.11 The Oxygen Challenge

The greatest obstacle remains oxygen.

Nitrogenase is oxygen sensitive.

Plant cells contain:

- chloroplasts,

- mitochondria,
- oxygen-rich environments.

Any functional nitroplast must therefore solve the oxygen problem.

Potential solutions include:

Protective Membranes

Oxygen Scavenging Systems

Temporal Regulation

Novel Nitrogenases

Light-Utilising Nitrogenases (LUN)

The solution may require entirely new biological designs.

14.12 Nitroplasts and Chloroplasts

A particularly intriguing possibility involves close integration between chloroplasts and nitroplasts.

Potential relationships include:

Energy Sharing

Chloroplasts provide ATP and reducing power.

Nitrogen Sharing

Nitroplasts provide ammonia.

Coordinated Regulation

Both systems respond to environmental conditions.

Such integration could create highly efficient photosynthetic nitrogen-fixing cells.

14.13 AI and Nitroplast Design

Artificial intelligence may play a crucial role.

Modern tools can assist with:

Protein Design

Pathway Optimization

Organelle Modeling

Genome Engineering

Metabolic Simulation

For the first time, scientists can computationally explore organelle architectures that have never existed in nature.

14.14 Nitroplasts and Agriculture

The agricultural implications are enormous.

Future crops possessing nitroplasts could potentially:

Fix Their Own Nitrogen

Reduce Fertilizer Requirements

Improve Sustainability

Increase Productivity

Reduce Environmental Pollution

Such crops would represent a major advance in global food security.

14.15 The Nitroplast Revolution

The Earth Clock Theory introduces the concept of:

The Nitroplast Revolution

This concept describes a future transition in which nitrogen fixation becomes integrated directly into higher plant cells.

Potential consequences include:

- redesigned agriculture,
- reduced industrial fertilizer dependence,
- improved ecosystem sustainability.

If achieved, the nitroplast revolution may become one of the most important biological transformations since the emergence of chloroplasts.

14.16 Beyond Earth

Nitroplasts may be particularly valuable beyond Earth.

Future settlements on:

- the Moon,
- Mars,
- orbital habitats,

will require highly efficient nutrient cycling.

Self-fertilizing plants could become essential components of extraterrestrial ecosystems.

Nitroplasts may therefore contribute to humanity's expansion into space.

14.17 The Earth Clock Perspective

Within the Earth Clock framework, organelles represent major milestones in biological evolution.

Mitochondria

Enabled complex eukaryotes.

Chloroplasts

Enabled photosynthetic ecosystems.

Nitroplasts

May enable a future generation of self-sustaining biological systems.

The progression reflects increasing metabolic integration and efficiency.

14.18 Future Directions

Major research questions remain:

Can nitrogenase function inside plant organelles?

Can oxygen sensitivity be overcome?

Can stable nitrogen-fixing compartments be engineered?

Can synthetic endosymbioses become permanent?

Could LUN systems accelerate nitroplast development?

These questions define a major frontier in modern biology.

Conclusions

The nitroplast concept represents a bold vision for the future of life.

Key conclusions include:

1. Mitochondria and chloroplasts arose through endosymbiosis.
2. Similar principles may eventually produce nitrogen-fixing organelles.
3. Natural examples provide important precedents.
4. Synthetic biology offers pathways toward artificial nitroplasts.
5. Oxygen sensitivity remains the principal obstacle.
6. Nitroplasts could transform agriculture and sustainability.
7. Future extraterrestrial ecosystems may benefit from nitroplast-based biology.

Within the Earth Clock framework, the nitroplast represents more than a technological goal.

It represents a potential new stage in biological evolution.

Mitochondria transformed energy.

Chloroplasts transformed sunlight.

Nitroplasts may transform nitrogen.

End of Chapter 14

Chapter 15

Light-Utilising Nitrogenase (LUN)

A New Paradigm for Biological Nitrogen Fixation

15.1 Introduction

Biological nitrogen fixation is widely recognized as one of the most important biochemical processes in the history of life. Yet despite more than a century of research, the central catalyst responsible for this process—nitrogenase—remains fundamentally constrained by two major limitations:

1. Extremely high energy consumption.
2. Extreme sensitivity to oxygen.

These limitations have shaped the ecological distribution of nitrogen-fixing organisms for billions of years and continue to hinder efforts to engineer nitrogen fixation into major crops.

The conventional molybdenum nitrogenase requires at least sixteen molecules of ATP to reduce one molecule of atmospheric nitrogen.

$$N_2 + 8H^+ + 8e^- + 16ATP \rightarrow 2NH_3 + H_2 + 16ADP + 16P_i$$

This energetic cost is enormous.

At the same time, oxygen rapidly inactivates nitrogenase, creating a paradox:

> The biosphere depends upon both nitrogen fixation and oxygenic photosynthesis, yet the enzyme responsible for nitrogen fixation is poisoned by oxygen.

This contradiction has persisted throughout evolutionary history.

The Light-Utilising Nitrogenase (LUN) hypothesis proposes a possible solution.

15.2 A Fundamental Evolutionary Question

Throughout Earth history, photosynthesis and nitrogen fixation have evolved in parallel.

Photosynthesis provides:

- Reducing power
- ATP
- Biomass
- Atmospheric oxygen

Nitrogen fixation provides:

- Bioavailable nitrogen
- Amino acids
- Proteins
- Genetic material

Together they support nearly all life on Earth.

Yet modern nitrogenase still relies on ATP-driven electron transfer mechanisms that may reflect ancient anaerobic conditions rather than optimal solutions for an oxygen-rich world.

This raises a fundamental question:

> Could evolution eventually produce a nitrogenase that directly utilizes light energy?

The LUN hypothesis begins from this question.

15.3 Lessons from Chlorophyll Biosynthesis

One of the strongest clues comes from chlorophyll biosynthesis.

Two distinct enzymes perform similar functions:

DPOR

Dark-operative protochlorophyllide oxidoreductase

Characteristics:

- Nitrogenase-like structure
- Fe-S clusters
- Oxygen sensitive
- Ancient origin

LPOR

Light-operative protochlorophyllide oxidoreductase

Characteristics:

- Uses light energy
- Oxygen tolerant
- Does not require ATP-driven electron transfer
- Dominant in higher plants

This evolutionary transition is remarkable.

Ancient Earth:

DPOR dominated.

Modern plants:

LPOR dominates.

The transition represents a shift from:

ATP-dependent catalysis

to

light-dependent catalysis.

The existence of LPOR demonstrates that evolution can replace ancient ATP-intensive systems with photochemical alternatives.

15.4 The LUN Hypothesis

The central proposition is simple:

> A future nitrogenase-like system may directly couple photochemical energy to nitrogen reduction.

Such an enzyme would represent a new class of catalyst:

Light-Utilising Nitrogenase (LUN)

Instead of relying exclusively on ATP hydrolysis, LUN could potentially exploit:

- Sunlight
- Photosynthetic electron transport
- Photochemical charge separation

to drive nitrogen reduction.

Conceptually:

Sunlight → Photosystem → Electron Transfer → Nitrogen Reduction

rather than:

ATP → Fe Protein → MoFe Protein → Nitrogen Reduction

15.5 Theoretical Advantages of LUN

A successful LUN system could potentially provide:

Reduced ATP Demand

One major limitation of biological nitrogen fixation is energy consumption.

A photochemical nitrogenase could bypass part of the ATP requirement.

This would dramatically improve nitrogen fixation efficiency.

Greater Compatibility with Photosynthesis

Current nitrogen fixation and oxygenic photosynthesis often require spatial or temporal separation.

Examples include:

- Root nodules
- Heterocysts
- Night-time nitrogen fixation

LUN could potentially operate more directly within photosynthetic environments.

Improved Agricultural Productivity

Future crops carrying LUN-like systems might:

- Require less fertilizer
- Produce higher yields
- Reduce environmental pollution

Planetary Sustainability

Reduced fertilizer dependence would lower:

- Greenhouse gas emissions
- Water pollution
- Energy consumption

globally.

15.6 Possible Evolutionary Pathways

Several theoretical evolutionary routes may lead toward LUN-like systems.

Route I

Nitrogenase Modification

Existing nitrogenases acquire photochemical modules through evolutionary innovation.

This route preserves:

- Nitrogen reduction chemistry

while replacing:

- ATP-dependent energy coupling.

Route II

DPOR-Derived Evolution

DPOR and nitrogenase share deep structural similarities.

A DPOR-like ancestor might evolve expanded catalytic capabilities.

Potential progression:

DPOR → Intermediate Systems → LUN

Route III

Synthetic Evolution

Human-directed evolution may accelerate processes that natural evolution would require millions of years to accomplish.

Tools include:

- Directed evolution
- AI-guided design
- Protein engineering
- Synthetic biology

This route is likely the fastest.

15.7 Structural Design Principles

A hypothetical LUN system would require several functional modules.

Module 1

Light-Harvesting Unit

Possible sources:

- Chlorophyll-binding proteins
- Photosystem fragments

- Synthetic chromophores

Purpose:

Capture photons.

Module 2

Electron Transfer Network

Function:

Transfer photogenerated electrons toward catalytic centers.

Possible components:

- Iron-sulfur clusters
- Flavins
- Quinones

Module 3

Catalytic Center

Potential candidates:

- Modified FeMoco
- Alternative metal cofactors
- Novel nitrogenase-like clusters

Function:

Reduce N_2.

Module 4

Oxygen Protection System

Essential for operation in photosynthetic environments.

Possible mechanisms:

- Protein shielding
- Rapid electron turnover
- Novel catalytic architectures

15.8 Artificial Intelligence and LUN

Recent advances in artificial intelligence have transformed protein design.

Modern tools include:

- AlphaFold
- RoseTTAFold
- RFdiffusion
- ProteinMPNN
- LigandMPNN

For the first time, scientists can computationally explore protein structures that have never existed in nature.

AI may help:

Identify Candidate Scaffolds

Search evolutionary space.

Predict Electron Pathways

Optimize charge transfer.

Design Novel Active Sites

Generate new catalytic architectures.

Improve Oxygen Tolerance

Predict mutations that stabilize catalytic centers.

AI therefore creates a realistic pathway toward experimental testing of LUN concepts.

15.9 Experimental Strategies

Potential laboratory approaches include:

Strategy A

Engineering nitrogenase inside chloroplasts.

Strategy B

Combining nitrogenase with photosystem-derived electron transfer systems.

Strategy C

Creating synthetic organelles (Nitroplasts).

Strategy D

Directed evolution under light-dependent conditions.

Strategy E

AI-guided de novo enzyme design.

15.10 Implications for the Earth Clock

Within the Earth Clock framework, nitrogen fixation was one of the earliest innovations of life.

Photosynthesis was another.

For billions of years these systems evolved independently yet cooperatively.

LUN represents a theoretical convergence.

If realized, it would unite:

- Nitrogen fixation
- Photosynthesis
- Synthetic biology
- Artificial intelligence

into a single evolutionary platform.

Such a transition could become one of the most important biological innovations since the emergence of oxygenic photosynthesis itself.

15.11 Beyond Earth

Future human civilizations may depend on biological systems capable of operating in extraterrestrial environments.

Potential applications include:

- Lunar agriculture
- Martian agriculture
- Closed ecological systems

- Deep-space habitats

LUN-like systems could provide self-sustaining nitrogen cycles in environments where industrial fertilizer production is impossible.

Thus the significance of LUN extends beyond agriculture.

It may become a foundational technology for the expansion of life beyond Earth.

15.12 Conclusions

The Light-Utilising Nitrogenase hypothesis proposes a new direction for biological nitrogen fixation research.

Although no naturally occurring LUN system has yet been discovered, evolutionary history demonstrates that biological innovation can repeatedly transform ancient biochemical mechanisms into new functional architectures.

The transition from DPOR to LPOR provides a compelling precedent.

Whether through natural evolution or synthetic biology, future nitrogen fixation systems may increasingly exploit light energy, improved oxygen tolerance, and advanced biological design.

If such systems emerge, they could fundamentally alter agriculture, sustainability, and the future trajectory of life on Earth.

> Nitrogen fixation transformed the ancient Earth.
> Light-utilising nitrogen fixation may help transform the future Earth.

End of Chapter 15

Chapter 16

Artificial Intelligence and Enzyme Design

From Natural Evolution to Intelligent Evolution

16.1 Introduction

For nearly four billion years, biological innovation was driven entirely by natural evolution.

Random mutations generated variation.

Natural selection filtered possibilities.

Successful solutions accumulated over time.

This process produced:

- nitrogenase,
- photosynthesis,
- chloroplasts,
- multicellular organisms,
- human intelligence.

Today, however, a new force has emerged.

Artificial Intelligence (AI) is beginning to influence biological discovery and molecular design.

For the first time in Earth's history:

Evolution is being joined by prediction.

Within the Earth Clock framework, this transition represents a major milestone.

Natural evolution created life.

Intelligent evolution may begin to redesign it.

16.2 The Limits of Natural Evolution

Natural evolution is extraordinarily powerful.

Yet it possesses limitations.

Randomness

Mutations occur largely without foresight.

Time

Complex innovations may require millions of years.

Local Optimization

Evolution often favors short-term solutions.

Historical Constraints

Existing structures limit future possibilities.

Despite these constraints, evolution produced astonishing biological complexity.

Artificial intelligence offers a way to accelerate and expand this process.

16.3 The Protein Design Challenge

Proteins are the fundamental molecular machines of life.

They perform:

- catalysis,
- signaling,
- transport,
- regulation,
- energy conversion.

A typical protein may contain hundreds of amino acids.

The number of possible sequences is enormous.

For a protein of 300 amino acids:

$$20^{300}$$

This number exceeds the number of atoms in the observable universe.

Traditional experimental approaches can explore only a tiny fraction of this space.

AI changes the situation dramatically.

16.4 The AlphaFold Revolution

One of the most significant breakthroughs in modern biology occurred with the development of:

AlphaFold

AlphaFold demonstrated that protein structures could be predicted with remarkable accuracy directly from amino acid sequences.

The impact was immediate.

Problems that once required years of experimental work could now be addressed computationally.

Consequences include:

Faster Discovery

Structural Insights

Functional Prediction

Accelerated Engineering

The AlphaFold revolution transformed structural biology.

16.5 Beyond Structure Prediction

Predicting structure was only the beginning.

New AI systems now address:

Protein Design

Enzyme Engineering

Molecular Interaction Prediction

Ligand Binding

Functional Optimization

AI is evolving from a predictive tool into a creative tool.

It is no longer limited to analyzing nature.

It can help invent new biological systems.

16.6 Protein Language Models

Protein sequences can be viewed as biological languages.

Just as language models learn grammar and meaning from text, protein language models learn patterns from biological sequences.

Examples include:

- ESM

- ProtT5
- ProGen
- EvoFormer-derived systems

These models can identify:

Conserved Motifs

Functional Relationships

Evolutionary Patterns

Novel Sequence Possibilities

Protein language models provide a powerful new framework for exploring biological design space.

16.7 RFdiffusion

A major advance in de novo protein design came with:

RFdiffusion

RFdiffusion generates entirely new protein structures. Rather than modifying existing proteins, it creates novel architectures.

Capabilities include:

Scaffold Design

Active Site Construction

Interface Engineering

Symmetric Assembly Design

For the first time, scientists can design proteins that have never existed in nature.

16.8 ProteinMPNN

Another transformative tool is:

ProteinMPNN

Given a protein structure, ProteinMPNN designs amino acid sequences likely to adopt that structure.

Applications include:

Stability Optimization

Functional Improvement

Sequence Exploration

Engineering Novel Catalysts

Together, RFdiffusion and ProteinMPNN create powerful workflows for enzyme engineering.

16.9 AI and Nitrogenase Engineering

Nitrogenase presents one of the most challenging targets in molecular biology.

Key difficulties include:

- complex metal cofactors,
- electron transfer pathways,
- oxygen sensitivity,
- ATP dependence.

AI may help address these challenges.

Potential applications include:

Active Site Redesign

Electron Pathway Optimization

Oxygen-Tolerance Engineering

Cofactor Environment Remodeling

Novel Nitrogenase Architectures

The possibility of redesigning nitrogen fixation is no longer purely theoretical.

16.10 AI and LUN Development

The Light-Utilising Nitrogenase (LUN) hypothesis depends heavily upon the discovery of new catalytic architectures.

AI offers several advantages.

Identification of Evolutionary Intermediates

Prediction of Photochemical Pathways

Design of Electron Transfer Networks

Integration of Light-Harvesting Modules

Creation of Novel Catalytic Centers

These capabilities may significantly accelerate exploration of LUN concepts.

16.11 AI and Synthetic Nitroplasts

Nitroplasts require integration of multiple biological systems.

Challenges include:

Metabolic Design

Organelle Architecture

Gene Regulation

Resource Allocation

Oxygen Control

AI can assist by simulating complex biological networks before experimental implementation.

This reduces development time and increases design efficiency.

16.12 Directed Evolution Meets Artificial Intelligence

Directed evolution has already transformed biotechnology.

AI enhances this process.

Traditional workflow:

Mutation

↓

Selection

↓

Optimization

AI-assisted workflow:

Prediction

↓

Design

↓

Construction

↓

Validation

↓

Optimization

This approach dramatically reduces search space.

The result is more efficient engineering.

16.13 Intelligent Evolution Theory

The Earth Clock Theory introduces:

Intelligent Evolution Theory

The theory proposes:

> Biological evolution is entering a new phase in which intelligent systems increasingly influence evolutionary outcomes.

For billions of years:

Evolution was blind.

Today:

Evolution can be guided.

This does not replace natural selection.

Instead, it supplements it.

Human intelligence and artificial intelligence become additional evolutionary forces.

16.14 A New Evolutionary Driver

Historically, major evolutionary drivers included:

Mutation

Selection

Drift

Environmental Change

The twenty-first century introduces a new factor:

Computational Design

For the first time, biological innovation may be accelerated by predictive algorithms.

This represents a profound transition.

16.15 AI as an Evolutionary Tool

AI does not create biology independently.

Instead, it amplifies biological understanding.

Its role resembles that of:

- microscopes,
- DNA sequencing,
- structural biology.

However, AI differs in one important respect.

It can generate new possibilities rather than merely observe existing ones.

This capability makes AI uniquely powerful.

16.16 Risks and Responsibilities

Powerful technologies require careful stewardship.

Potential concerns include:

Unintended Ecological Effects

Biosecurity Risks

Genetic Instability

Ethical Challenges

Unequal Access

Responsible innovation remains essential.

The Earth Clock framework emphasizes that long-term planetary consequences must be considered.

16.17 AI and the Future of Agriculture

AI-guided biological engineering may transform agriculture.

Potential outcomes include:

Self-Fertilizing Crops

Climate-Resilient Plants

Reduced Fertilizer Dependence

Improved Sustainability

Precision Agriculture

The combination of AI and nitrogen fixation research may reshape global food production.

16.18 AI and the Future Earth

Within the Earth Clock framework:

Nitrogen Fixation

Expanded biological productivity.

Photosynthesis

Transformed planetary energy flow.

Artificial Intelligence

May transform biological innovation itself.

This possibility marks a new chapter in Earth's evolutionary history.

Conclusions

Artificial intelligence is becoming a major force in biological research.

Key conclusions include:

1. AI dramatically expands protein design capabilities.
2. AlphaFold transformed structural biology.
3. RFdiffusion and ProteinMPNN enable de novo protein engineering.
4. Nitrogenase engineering may benefit substantially from AI.
5. LUN and nitroplast development may depend on AI-assisted design.
6. Biological evolution is increasingly influenced by intelligent systems.
7. AI may become a significant evolutionary force during the coming centuries.

Within the Earth Clock framework, artificial intelligence represents more than a technological advance.

It represents a new relationship between knowledge and evolution.

Evolution created intelligence.

Intelligence created artificial intelligence.

Artificial intelligence may help guide the next stage of evolution.

End of Chapter 16

Chapter 17

Self-Fertilising Crops

The Next Agricultural Revolution

17.1 Introduction

Agriculture is one of humanity's greatest achievements.

The domestication of plants transformed:

- human societies,
- economies,
- civilizations,
- ecosystems.

Yet modern agriculture remains fundamentally dependent upon external nitrogen inputs.

Despite thousands of years of agricultural development, most major crops cannot directly utilize atmospheric nitrogen.

Instead they depend upon:

- soil nitrogen,
- microbial activity,
- industrial fertilizers.

This dependence creates environmental and economic challenges.

Within the Earth Clock framework, the development of self-fertilising crops represents a potential new stage in agricultural evolution.

17.2 The Nitrogen Dependency Problem

Nitrogen is often the most limiting nutrient in crop production.

Modern agriculture compensates through fertilizer application.

Global fertilizer production exceeds hundreds of millions of tons annually.

Benefits include:

Increased Yield

Improved Food Security

Population Support

However, costs include:

Water Pollution

Greenhouse Gas Emissions

Soil Degradation

Energy Consumption

The challenge is not merely producing nitrogen.

The challenge is producing nitrogen sustainably.

17.3 The Dream of Self-Fertilising Plants

For over a century, scientists have pursued a remarkable goal:

Crops capable of generating their own nitrogen.

Such plants would:

- reduce fertilizer requirements,
- improve sustainability,
- increase agricultural resilience.

The concept appears simple.

The implementation is extraordinarily difficult.

Nitrogen fixation requires complex biological machinery.

Yet the potential rewards justify continued effort.

17.4 Why Legumes Are Different

Nature already provides an example.

Legumes such as:

- soybean,
- pea,
- clover,
- alfalfa,

form symbiotic relationships with nitrogen-fixing bacteria.

Within root nodules:

Rhizobia

convert atmospheric nitrogen into ammonia.

The plant supplies:

- carbohydrates,
- energy,
- protective environments.

The bacteria supply:

- fixed nitrogen.

This partnership demonstrates that biological nitrogen fixation can support highly productive agriculture.

17.5 Why Cereals Cannot Fix Nitrogen

Major cereal crops include:

Wheat

Rice

Maize

Barley

Together they provide much of humanity's food supply.

However, cereals lack the sophisticated symbiotic systems found in legumes.

Consequently they rely heavily on fertilizers.

The challenge is introducing nitrogen-fixing capability without disrupting normal plant function.

17.6 Strategy I:

Improved Symbiosis

One approach involves engineering improved associations between cereals and nitrogen-fixing microorganisms.

Potential methods include:

Enhanced Root Colonization

Synthetic Signaling Systems

Engineered Diazotrophs

Improved Nutrient Exchange

Rather than modifying plants extensively, this strategy enhances microbial partnerships.

17.7 Strategy II:

Endophytic Nitrogen Fixation

Certain bacteria naturally inhabit internal plant tissues.

These organisms are known as:

Endophytes

Some possess nitrogen-fixing capability.

Advantages include:

- close proximity to plant cells,
- reduced nitrogen loss,
- protection from environmental fluctuations.

Engineering endophytic systems may provide practical routes toward self-fertilising crops.

17.8 Strategy III:

Direct Nitrogenase Engineering

A more ambitious approach involves transferring nitrogen-fixation machinery directly into plants.

Requirements include:

nif Genes

Electron Transfer Systems

Cofactor Assembly Pathways

Regulatory Networks

Oxygen Protection Mechanisms

This strategy seeks to transform plants themselves into nitrogen-fixing organisms.

The complexity remains enormous.

17.9 Strategy IV: Chloroplast-Based Nitrogen Fixation

Chloroplasts offer unique opportunities.

Advantages include:

Bacterial Evolutionary Origins

High Gene Expression

Energy Availability

Compartmentalization

Researchers have proposed chloroplasts as platforms for future nitrogen fixation systems.

The challenge remains oxygen management.

17.10 Strategy V:

Synthetic Nitroplasts

Nitroplasts represent perhaps the most ambitious strategy.

Instead of modifying existing cellular compartments, scientists could create entirely new organelles dedicated to nitrogen fixation.

Potential benefits include:

Controlled Environments

Metabolic Integration

Oxygen Regulation

Efficient Nitrogen Delivery

Nitroplasts may ultimately provide the most comprehensive solution.

17.11 Strategy VI:

Light-Utilising Nitrogenase

The LUN hypothesis introduces another possibility. Photochemical nitrogen fixation could:

- reduce ATP requirements,
- improve integration with photosynthesis,
- increase efficiency.

Future self-fertilising crops may depend upon combinations of:

LUN

Nitroplasts

AI-designed proteins

Synthetic biology

The most successful solutions may emerge from integrated approaches.

17.12 Artificial Intelligence and Crop Design

Artificial intelligence is accelerating agricultural biotechnology.

Applications include:

Gene Discovery

Pathway Optimization

Protein Engineering

Trait Prediction

Evolutionary Modeling

AI may help identify solutions that natural evolution never explored.

This capability could dramatically accelerate crop development.

17.13 The Agricultural Evolution Theory

The Earth Clock Theory introduces:

Agricultural Evolution Theory

The theory proposes:

> Agriculture evolves through successive increases in biological autonomy.

Historical sequence:

Wild Plants

↓

Domesticated Crops

↓

Fertilizer-Dependent Agriculture

↓

Precision Agriculture

↓

Self-Fertilising Crops

Each stage reduces dependence on external limitations.

Self-fertilising crops represent the next logical step.

17.14 Environmental Benefits

Potential environmental benefits include:

Reduced Fertilizer Production

Lower Energy Consumption

Reduced Nitrogen Runoff

Improved Water Quality

Reduced Greenhouse Gas Emissions

Such improvements would affect ecosystems globally.

17.15 Food Security Benefits

Global food demand continues to rise.

Self-fertilising crops could contribute to:

Stable Yields

Reduced Production Costs

Improved Resource Efficiency

Greater Agricultural Resilience

These benefits are particularly important for regions with limited access to fertilizers.

17.16 Challenges and Risks

Significant obstacles remain.

Nitrogenase Complexity

Oxygen Sensitivity

Metabolic Costs

Ecological Stability

Regulatory Concerns

Future success requires careful scientific and societal evaluation.

17.17 Self-Fertilising Crops and Civilization

Agricultural innovation has repeatedly reshaped civilization.

Examples include:

Domestication

Irrigation

Mechanization

Fertilizers

Genetic Improvement

Self-fertilising crops may become the next major agricultural transition.

Their impact could rival earlier revolutions.

17.18 The Earth Clock Perspective

Within the Earth Clock framework:

Nitrogen Fixation

Enabled biological expansion.

Agriculture

Enabled civilization.

Self-Fertilising Crops

May enable sustainable civilization.

The progression reflects increasing mastery of biological resources.

17.19 Toward a Nitrogen-Neutral Agriculture

Future agriculture may move toward:

Reduced External Inputs

Closed Nutrient Cycles

Enhanced Biological Function

Ecosystem Integration

Self-fertilising crops could become central components of this vision.

Conclusions

Self-fertilising crops represent one of the most important goals in modern agriculture.

Key conclusions include:

1. Modern agriculture remains highly dependent on nitrogen fertilizers.
2. Biological nitrogen fixation offers sustainable alternatives.
3. Multiple engineering strategies are being explored.
4. Nitroplasts and LUN systems may provide future solutions.
5. Artificial intelligence is accelerating crop innovation.

6. Self-fertilising crops could transform global food systems.
7. Agricultural evolution is moving toward greater biological autonomy.

Within the Earth Clock framework, self-fertilising crops represent more than an agricultural technology. They represent a potential evolutionary milestone linking nitrogen fixation, synthetic biology and sustainable civilization.

Nitrogen fixation enabled life.

Agriculture enabled civilization.

Self-fertilising crops may help secure the future of both.

End of Chapter 17

Chapter 18

Nitrogen Fixation Beyond Earth

Cosmic Agriculture and the Future Expansion of Life

18.1 Introduction

For billions of years, biological nitrogen fixation has operated exclusively on Earth.

Yet humanity is rapidly entering a new era.

Future generations may establish permanent settlements on:

- the Moon,
- Mars,
- orbital habitats,
- artificial space colonies.

These environments present extraordinary challenges.

Among them, one of the most fundamental is:

Nitrogen Availability

Life cannot exist without nitrogen.

Proteins require nitrogen.

DNA requires nitrogen.

Ecosystems require nitrogen.

Thus any long-term extraterrestrial civilization must solve the nitrogen problem.

Within the Earth Clock framework, the expansion of life beyond Earth represents the next great frontier of biological evolution.

18.2 The Universal Importance of Nitrogen

The importance of nitrogen is not uniquely terrestrial.

Any carbon-based life form relying upon:

- amino acids,
- nucleic acids,
- cellular metabolism,

will require nitrogen.

Consequently, nitrogen cycles may represent universal features of living planets.

The Earth Clock Theory proposes:

> Wherever life becomes complex, nitrogen acquisition becomes essential.

This principle may apply throughout the Universe.

18.3 Why Space Agriculture Matters

Long-duration space missions require sustainable food production.

Transporting food indefinitely from Earth is impractical.

Future settlements must become increasingly self-sufficient.

Requirements include:

Food Production

Water Recycling

Oxygen Generation

Waste Processing

Nutrient Cycling

Agriculture therefore becomes a central component of space colonization.

18.4 The Nitrogen Challenge in Space

On Earth, nitrogen cycles naturally through:

- atmosphere,
- oceans,

- soils,
- ecosystems.

In space habitats, these natural reservoirs are absent or limited.

Every atom of nitrogen becomes valuable.

Losses cannot be easily replaced.

Consequently:

Nitrogen recycling and nitrogen fixation become critical technologies.

Without them, long-term habitation becomes difficult.

18.5 Lunar Agriculture

The Moon presents unique challenges.

Characteristics include:

No Atmosphere

Extreme Temperature Variation

Limited Water Availability

High Radiation

Reduced Gravity

Lunar agriculture will likely occur within controlled environments.

Potential systems include:

- greenhouses,
- underground habitats,
- closed ecological systems.

Nitrogen must be carefully managed and recycled.

18.6 Martian Agriculture

Mars offers greater opportunities.

Advantages include:

Atmospheric Presence

Water Ice

Day–Night Cycles

Geological Resources

However, significant challenges remain.

The Martian atmosphere consists primarily of:

- carbon dioxide,
- trace nitrogen,
- trace oxygen.

Available nitrogen is limited.

Future Martian agriculture may therefore depend heavily upon biological nitrogen fixation.

18.7 Closed Ecological Systems

One of the most important concepts in space biology is:

Closed Ecological Systems

These systems recycle:

- carbon,
- oxygen,
- water,
- nutrients.

A simplified cycle appears below.

Plants

↓

Food

↓

Humans

↓

Waste

↓

Microorganisms

↓

Nutrients

↓

Plants

Nitrogen cycling is one of the most important components of such systems.

Without efficient nitrogen recycling, ecological stability cannot be maintained.

18.8 Biological Nitrogen Fixation in Space

Nitrogen-fixing organisms offer several advantages.

They can:

Produce Ammonia

Support Plant Growth

Improve Sustainability

Reduce Dependence on External Inputs

Potential candidates include:

- cyanobacteria,
- diazotrophic bacteria,
- engineered microorganisms.

These organisms may become essential components of extraterrestrial ecosystems.

18.9 Self-Fertilising Crops Beyond Earth

The value of self-fertilising crops increases dramatically in space.

Benefits include:

Reduced Fertilizer Transport

Greater Resource Efficiency

Simplified Agricultural Systems

Improved Reliability

Future lunar and Martian agriculture may depend heavily upon crops possessing enhanced nitrogen-fixing capabilities.

18.10 Nitroplasts in Space Agriculture

Nitroplasts could be particularly valuable beyond Earth.

Potential advantages include:

Autonomous Nitrogen Acquisition

Efficient Nutrient Cycling

Reduced Infrastructure Requirements

Increased Agricultural Independence

Space environments may provide strong incentives for developing nitroplast-based biology.

18.11 LUN and Extraterrestrial Agriculture

The Light-Utilising Nitrogenase (LUN) hypothesis may have unique relevance in space.

Solar energy is abundant.

Fertilizer production is difficult.

Photochemical nitrogen fixation could therefore become highly attractive.

Potential benefits include:

Reduced ATP Requirements

Greater Energy Efficiency

Improved Crop Productivity

Enhanced Sustainability

LUN-based systems may become important components of future extraterrestrial agriculture.

18.12 Astrobiology and Nitrogen

Nitrogen also plays a central role in astrobiology.

Scientists searching for extraterrestrial life often examine:

Atmospheric Composition

Chemical Disequilibrium

Nutrient Availability

Habitability Indicators

Nitrogen is frequently considered a key element in these analyses.

The Earth Clock Theory predicts that nitrogen acquisition systems may be common features of inhabited worlds.

18.13 The Cosmic Nitrogen Cycle

The Earth Clock framework extends the nitrogen cycle beyond Earth.

A generalized cosmic nitrogen cycle may involve:

Atmosphere

↓

Nitrogen Fixation

↓

Biomass

↓

Food Webs

↓

Recycling

↓

Atmosphere

Although specific mechanisms may differ among planets, the underlying principles remain similar.

Life requires continuous access to nitrogen.

18.14 Cosmic Agriculture Theory

The Earth Clock Theory introduces:

Cosmic Agriculture Theory

The theory proposes:

> Agriculture will eventually expand beyond Earth and become a planetary-scale and interplanetary-scale biological process.

Historical sequence:

Natural Ecosystems

↓

Human Agriculture

↓

Industrial Agriculture

↓

Self-Fertilising Agriculture

↓

Space Agriculture

↓

Interplanetary Agriculture

Each stage expands the domain within which life manages nutrients and energy.

18.15 Terraforming and Nitrogen

Future discussions of planetary engineering often focus on:

- oxygen,
- temperature,
- water.

Nitrogen is equally important.

A functioning biosphere requires:

Nitrogen Sources

Nitrogen Cycling

Nitrogen Retention

Without these processes, ecological systems remain unstable.

Nitrogen fixation may therefore play important roles in future terraforming strategies.

18.16 Humanity as an Evolutionary Force

The expansion of life beyond Earth introduces a new evolutionary phenomenon.

For the first time:

Life becomes capable of intentionally spreading itself across planets.

Humanity acts as:

Explorer

Engineer

Ecological Designer

Evolutionary Agent

This transition has profound implications for the future of life.

18.17 The Earth Clock and the Cosmic Future

The Earth Clock Theory views the expansion of life into space as a continuation of biological evolution.

Past:

Life emerged.

↓

Nitrogen fixation evolved.

↓

Photosynthesis transformed Earth.

↓

Complex life appeared.

↓

Civilization emerged.

↓

Future:

Life expands beyond Earth.

This progression reflects increasing biological reach across larger spatial scales.

18.18 A Universe Becoming Biological

Earth currently represents the only known living world.

Yet if life spreads beyond Earth, the consequences could be extraordinary.

Biological systems may eventually inhabit:

- moons,
- planets,
- artificial habitats,
- interstellar environments.

Nitrogen fixation could accompany this expansion.

The same process that helped transform Earth may help sustain future worlds.

Conclusions

Nitrogen fixation will remain essential as humanity expands into space.

Key conclusions include:

1. Nitrogen is indispensable for all known forms of life.

2. Space agriculture requires efficient nutrient cycling.
3. Biological nitrogen fixation provides sustainable solutions.
4. Self-fertilising crops may become increasingly important beyond Earth.
5. Nitroplasts and LUN systems could support extraterrestrial ecosystems.
6. Nitrogen acquisition may be a universal feature of living worlds.
7. The future expansion of life will likely depend upon effective nitrogen management.

Within the Earth Clock framework, nitrogen fixation is not merely a process that shaped Earth's past.

It may become one of the technologies that shapes life's cosmic future.

Nitrogen fixation helped life conquer Earth.

It may one day help life reach the stars.

End of Chapter 18

Chapter 19

The Earth Clock Theory

Biological Nitrogen Fixation as a Central Constraint on the Evolution of Complex Life

19.1 Introduction

The history of life is often described through a sequence of major evolutionary innovations:

- The origin of life
- Photosynthesis
- Eukaryotic cells
- Multicellularity
- Animals
- Plants
- Humans

These events are usually treated as independent milestones.

However, a deeper examination suggests that they may be linked by a common underlying constraint:

the availability of biologically usable nitrogen.

The Earth Clock Theory proposes that biological nitrogen fixation was not merely one evolutionary innovation among many.

Instead:

> Nitrogen fixation acted as a central planetary constraint that influenced the timing, scale and complexity of biological evolution throughout Earth history.

In this view, the evolution of complex life cannot be fully understood without understanding the history of nitrogen fixation.

19.2 The Earth as a Biological Clock

The Earth Clock Theory begins with a simple observation.

Earth has:

- a beginning,
- a developmental trajectory,
- and an end.

Earth formed approximately 4.54 billion years ago.

Astronomical models predict that increasing solar luminosity will eventually render Earth uninhabitable.

Thus Earth's biological lifespan is finite.

The Earth Clock framework treats Earth as a long-lived evolving system whose total lifespan is approximately:

$$10^{10}\ \text{years}$$

Within this timescale, major biological innovations appear as sequential events occupying specific positions on a planetary timeline.

19.3 The Nitrogen Constraint

Every known living organism requires nitrogen.

Nitrogen is a fundamental component of:

- amino acids
- proteins
- nucleic acids
- ATP
- chlorophyll

Without fixed nitrogen:

- cells cannot divide
- genomes cannot expand
- complexity cannot increase

Yet atmospheric nitrogen is chemically inaccessible to most organisms.

The challenge originates from the extraordinary stability of molecular nitrogen.

$$N\equiv N$$

The Earth therefore faced a paradox:

An atmosphere rich in nitrogen,

but a biosphere poor in usable nitrogen.

The solution emerged through biological nitrogen fixation.

19.4 The First Planetary Bottleneck

The Earth Clock Theory identifies nitrogen availability as one of the earliest bottlenecks in biological evolution.

Before nitrogen fixation evolved:

- biomass was limited
- productivity was restricted

- ecosystem expansion was constrained

Life may have persisted locally.

However, large-scale biological diversification would have been difficult.

The emergence of nitrogen fixation removed this bottleneck.

For the first time:

Atmospheric nitrogen became biologically accessible.

The consequences were global.

19.5 Nitrogen Fixation Before Photosynthesis?

One of the most intriguing questions in evolutionary biology concerns the relative timing of nitrogen fixation and photosynthesis.

Many phylogenetic studies suggest that nitrogen fixation is extremely ancient.

Nitrogenase-like enzymes may have originated very early in cellular evolution.

If true:

Nitrogen fixation may have preceded oxygenic photosynthesis.

Under the Earth Clock Theory:

This possibility is highly significant.

Because it implies that:

> the nitrogen economy of life was established before the oxygen economy of life.

In other words:

Nitrogen fixation may have laid the biological foundation upon which later photosynthetic expansion was built.

19.6 Photosynthesis and Nitrogen Fixation: The Dual Engine Hypothesis

The Earth Clock Theory proposes that life on Earth has been powered by two fundamental biochemical engines.

Engine One

Photosynthesis

Provides:

- energy

- reducing power
- oxygen
- biomass

Engine Two

Nitrogen Fixation

Provides:

- amino acids
- proteins
- genetic material
- cellular growth capacity

Neither system alone is sufficient.

Photosynthesis without nitrogen fixation eventually encounters nutrient limitation.

Nitrogen fixation without photosynthesis lacks sufficient energy input.

Together they create the modern biosphere.

This relationship may represent one of the most important cooperative interactions in the history of life.

19.7 Oxygen:

Evolution's Greatest Opportunity and Challenge

The rise of oxygen transformed Earth.

The Great Oxidation Event enabled:

- aerobic respiration
- larger genomes
- greater metabolic efficiency
- multicellular organisms

However oxygen also introduced a new problem.

Nitrogenase is highly oxygen sensitive.

Thus evolution faced a contradiction.

The molecule enabling complex life simultaneously threatened one of the enzymes essential for sustaining life.

Throughout Earth's history, organisms developed numerous strategies to solve this conflict:

- heterocysts
- root nodules
- temporal separation
- oxygen scavenging systems

These adaptations illustrate the continuing importance of nitrogen fixation even after oxygen became abundant.

19.8 Complex Life as a Nitrogen-Dependent Phenomenon

The Earth Clock Theory proposes:

> Complex life is fundamentally a nitrogen-dependent phenomenon.

As organisms become more complex:

- genome size increases
- protein diversity expands
- developmental programs become more elaborate

All of these processes require nitrogen.

Consequently:

The evolution of complexity may depend not only on energy availability,

but also on nitrogen availability.

This concept can be summarized as:

Energy Enables

Nitrogen Builds

Photosynthesis provides energy.

Nitrogen fixation provides biological construction material.

Together they enable complexity.

19.9 The Nitrogen Threshold Hypothesis

The Earth Clock Theory introduces a new concept:

The Nitrogen Threshold Hypothesis

Major evolutionary transitions occur only after sufficient fixed nitrogen becomes available.

Possible examples include:

- expansion of microbial ecosystems
- emergence of eukaryotes
- diversification of plants
- evolution of terrestrial ecosystems

Under this hypothesis:

Nitrogen availability acts as a regulator of evolutionary potential.

Not every evolutionary innovation is directly caused by nitrogen fixation.

However nitrogen fixation determines how much biological complexity the Earth system can support.

19.10 Humanity:

A Late Arrival on the Earth Clock

From a planetary perspective, humanity appeared extremely recently.

If Earth's biological history is compressed into a single century:

- microbial life occupies most of the timeline
- photosynthesis dominates the middle period
- humans appear during the final moments

This observation leads to an important realization.

Human civilization depends entirely upon biological systems that evolved billions of years earlier.

Every harvest,

every ecosystem,

every food chain,

ultimately traces back to ancient microbial innovations.

Among these innovations, nitrogen fixation remains indispensable.

19.11 The Anthropocene Nitrogen Revolution

Humanity has recently altered the nitrogen cycle on a planetary scale.

The development of the Haber–Bosch process dramatically increased nitrogen fertilizer production.

N_2+3H_2\rightarrow2NH_3

For the first time in Earth's history:

Industrial nitrogen fixation rivaled biological nitrogen fixation.

This achievement supported:

- population growth
- agricultural expansion
- global food security

Yet it also created new challenges:

- eutrophication
- greenhouse gas emissions
- ecosystem disruption

Humanity has become an active participant in the Earth Clock.

19.12 The Next Evolutionary Transition

The Earth Clock Theory predicts that future biological evolution may increasingly involve engineered nitrogen fixation.

Possible developments include:

Self-fertilising cereals

Synthetic nitroplasts

Light-utilising nitrogenases

AI-designed enzymes

Artificial ecosystems

For billions of years evolution proceeded through natural selection.

For the first time:

Life may become capable of intentionally redesigning its own nitrogen economy.

19.13 Earth Clock Prediction I

The Nitrogen Future

Nitrogen fixation will remain essential throughout the future history of Earth.

Regardless of technological progress:

Proteins will still require nitrogen.

Genomes will still require nitrogen.

Life will still require nitrogen.

Therefore:

Any sustainable future civilization must maintain an efficient nitrogen cycle.

19.14 Earth Clock Prediction II

Synthetic Evolution

Future biological innovation may increasingly occur through engineered evolution rather than natural evolution.

Artificial intelligence may dramatically accelerate:

- protein design
- metabolic engineering
- synthetic organelle development

The pace of innovation may exceed anything observed during natural evolutionary history.

19.15 Earth Clock Prediction III

Life Beyond Earth

If humanity establishes permanent settlements beyond Earth:

Nitrogen fixation will become even more important.

Future habitats on:

- the Moon
- Mars
- orbital stations

will require self-sustaining nitrogen cycles.

Nitrogen-fixing biological systems may therefore accompany humanity wherever life expands.

19.16 A Unified View of Life

The Earth Clock Theory ultimately seeks to connect multiple disciplines.

It integrates:

- evolutionary biology
- microbiology
- geochemistry

- photosynthesis research
- nitrogen fixation research
- synthetic biology
- astrobiology

into a single temporal framework.

The central proposition is simple:

> Biological nitrogen fixation was not merely a biochemical innovation.

It was one of the principal mechanisms that allowed Earth to become a living planet.

19.17 Conclusions

The Earth Clock Theory proposes that biological nitrogen fixation has acted as a major evolutionary constraint throughout Earth's history.

The theory emphasizes three key ideas:

First

Nitrogen fixation was among the earliest transformative innovations of life.

Second

Photosynthesis and nitrogen fixation function as the dual biochemical engines of the biosphere.

Third

Future biological evolution may increasingly depend upon engineered forms of nitrogen fixation.

Within the Earth Clock framework, the history of life can be viewed as a long sequence of innovations built upon the availability of fixed nitrogen.

The emergence of nitrogen fixation helped create the biosphere of the past.

Its future evolution may help shape the biosphere of tomorrow.

The Earth Clock continues to tick.

Nitrogen fixation remains one of its most important gears.

End of Chapter 19

End of *The Earth Clock I: Biological Nitrogen Fixation and the Evolution of Complex Life* (2026 International Expanded Edition)

Appendix

Appendix A

Figure Captions

Figures 1–50

Figure 1

The Earth Clock Framework

A conceptual representation of Earth as a finite-duration biological system spanning approximately 10 billion years. The diagram illustrates Earth's formation, biological evolution, present state and projected future, emphasizing that life unfolds within a constrained planetary timescale.

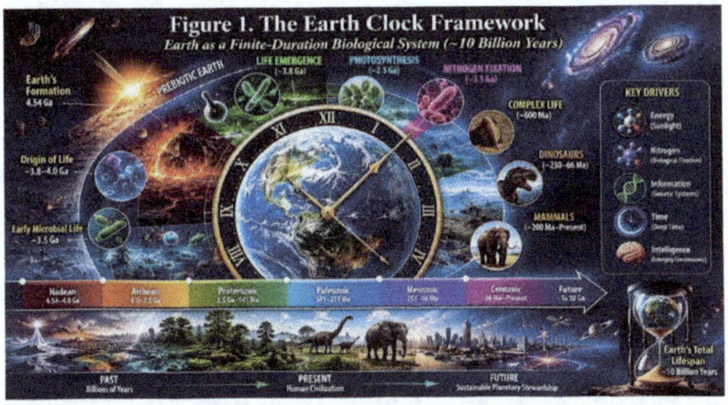

Figure 2

Cosmic Address of Earth

Hierarchical location of Earth within the Universe:

Human → Earth → Solar System → Orion Arm → Milky Way Galaxy → Local Group → Virgo Supercluster → Observable Universe.

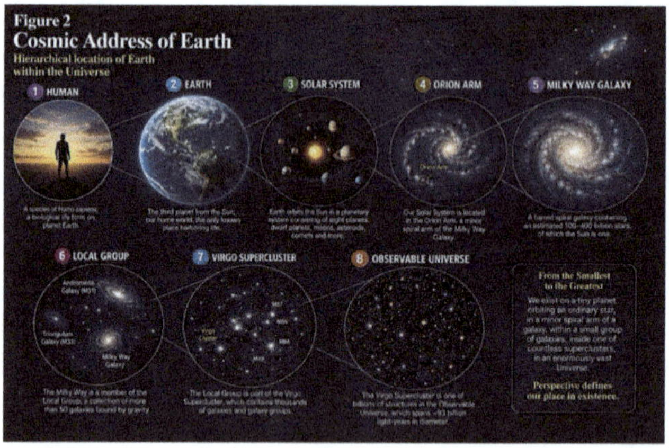

Figure 3

The Earth Clock Timeline

Simplified timeline of Earth's history showing major biological innovations:

Origin of Life → Nitrogen Fixation → Photosynthesis → Oxygenation → Eukaryotes → Plants → Animals → Humans.

Figure 4

The Nitrogen Crisis of Early Earth

Illustration of abundant atmospheric nitrogen but limited biological accessibility due to the stability of the N≡N triple bond.

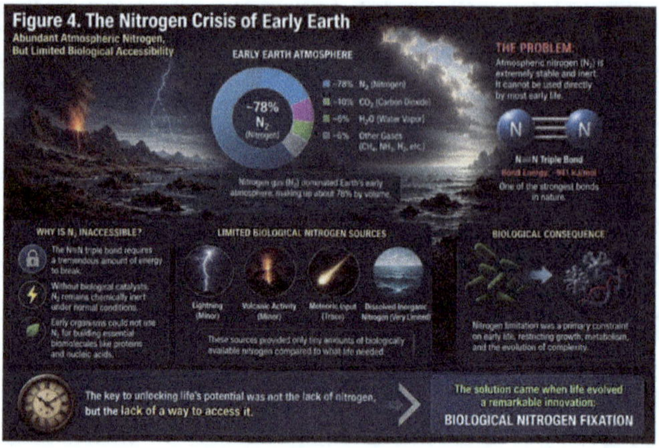

Figure 5

Abiotic Nitrogen Sources Before Biological Nitrogen Fixation

Potential prebiotic nitrogen inputs including:

* Lightning
* Volcanism
* Hydrothermal Systems
* Meteorite Delivery

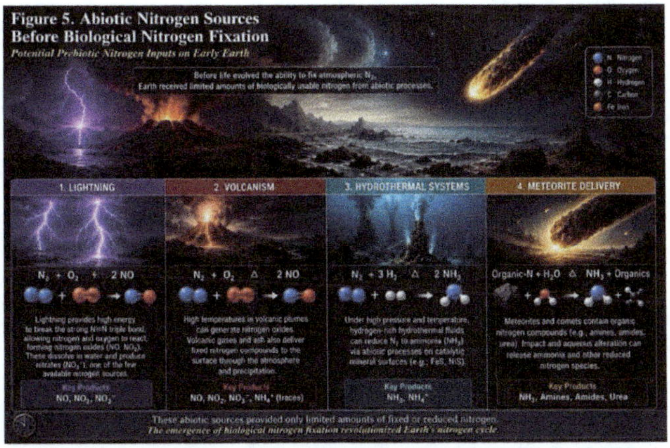

Figure 6

The Nitrogen Bottleneck Hypothesis

Conceptual model showing how limited fixed nitrogen constrained biomass accumulation and evolutionary diversification during early Earth history.

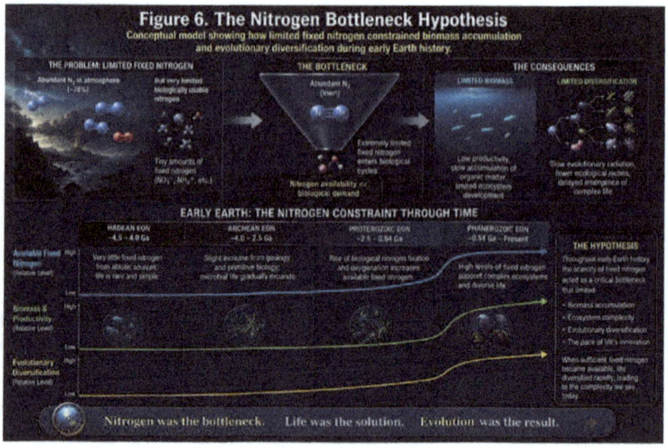

Figure 7

Emergence of Nitrogenase

Evolutionary reconstruction depicting the appearance of nitrogen fixation as one of Earth's earliest major biological innovations.

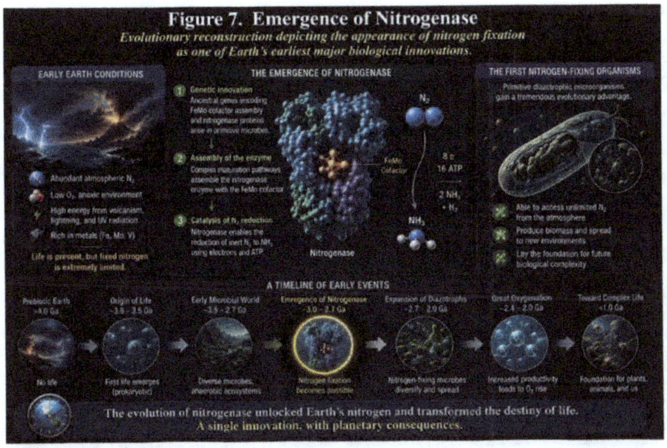

Figure 8

The First Global Nitrogen Cycle

Schematic showing nitrogen movement among atmosphere, microbes, oceans and sediments following the emergence of biological nitrogen fixation.

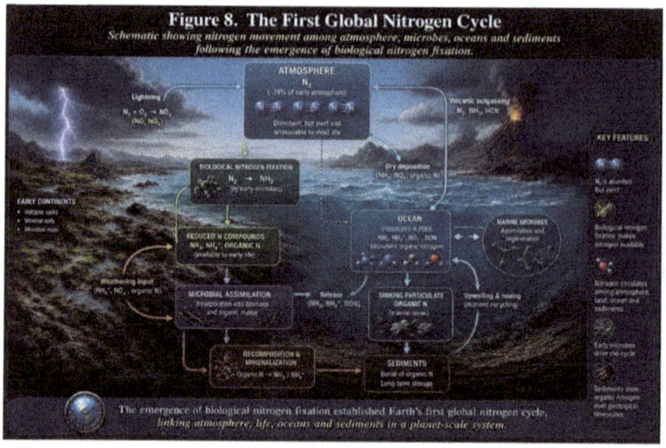

Figure 9

Nitrogenase Catalytic Reaction

Overall biochemical reaction catalyzed by nitrogenase.

Nitrogen gas is converted into ammonia through ATP-dependent electron transfer.

Figure 10

Structure of Classical Nitrogenase

Organization of:

* Fe Protein
* MoFe Protein
* P Cluster
* FeMoco

showing electron flow pathways.

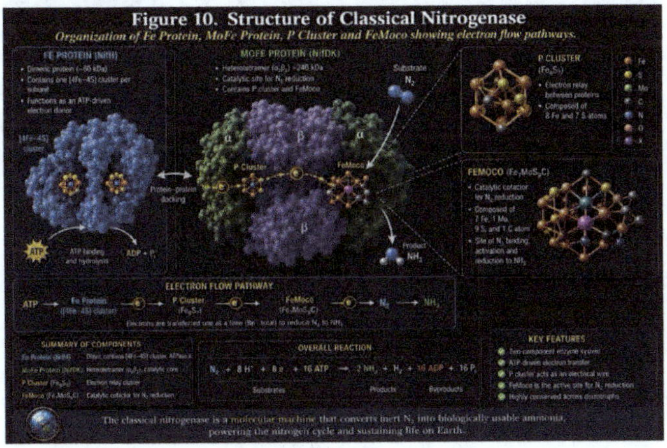

Figure 11

FeMoco Architecture

Detailed structure of the iron–molybdenum cofactor illustrating one of nature's most sophisticated catalytic metal clusters.

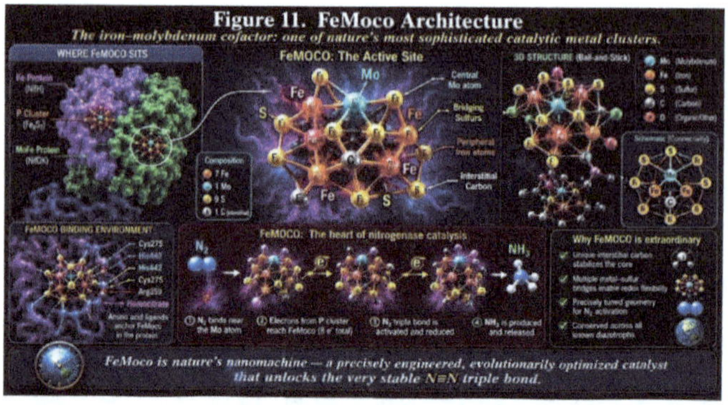

Figure 12

Nitrogenase Catalytic Cycle

Stepwise illustration of ATP binding, electron transfer, substrate reduction and product release during nitrogen fixation.

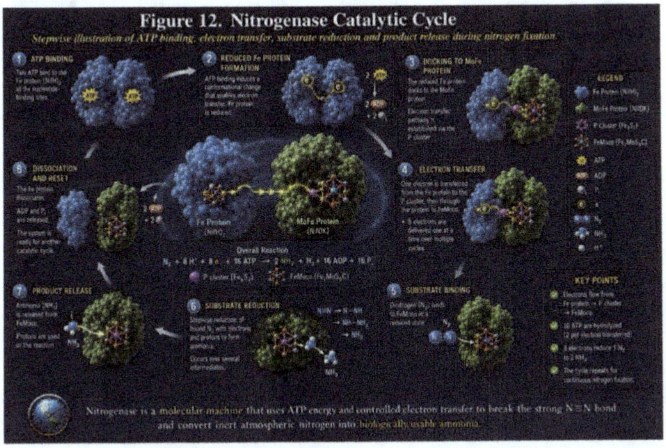

Figure 13

Evolution of Nitrogenase Systems

Relationship among:

* Fe-only Nitrogenase
* Vanadium Nitrogenase
* Molybdenum Nitrogenase

under changing planetary geochemistry.

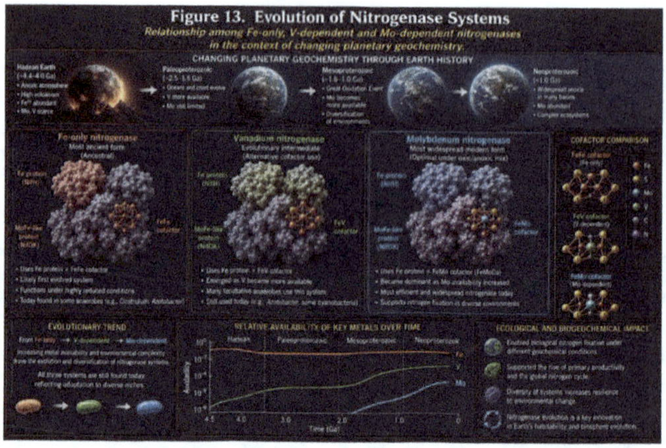

Figure 14

The Metal Age Hypothesis

Proposed connection between ocean chemistry and nitrogenase diversification during Earth's early evolution.

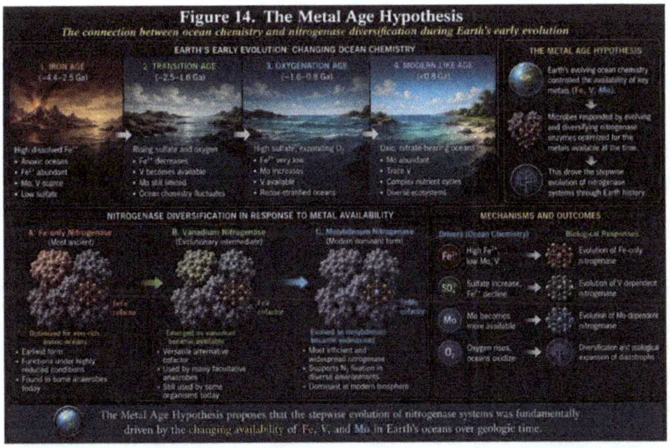

Figure 15

Nitrogenase Family Tree

Phylogenetic reconstruction of major nitrogenase systems and related enzyme families.

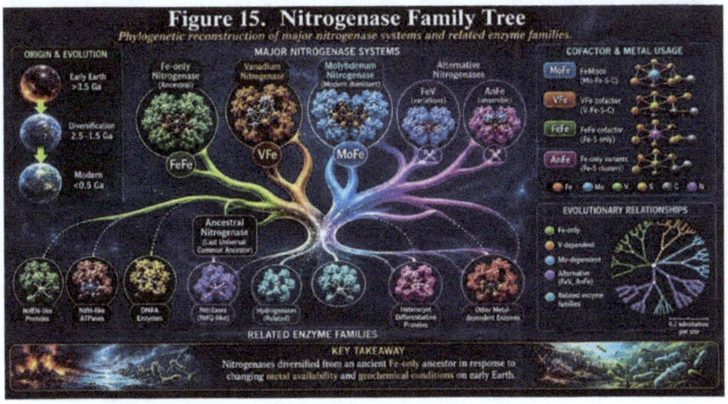

Figure 16

Alternative Nitrogenases

Comparison of:

* nif
* vnf
* anf

systems including catalytic properties and environmental distributions.

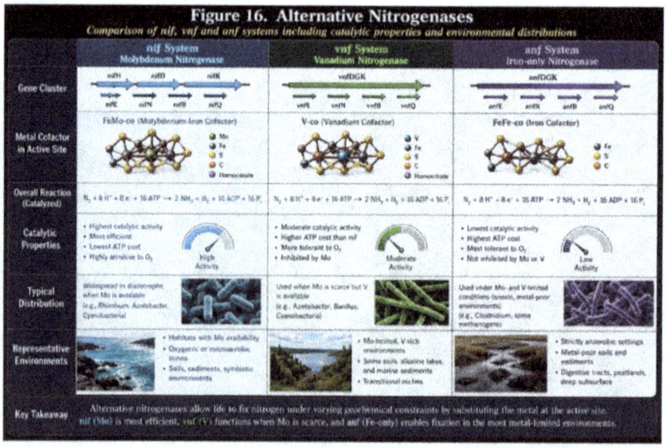

Figure 17

Catalytic Diversity of Alternative Nitrogenases

Illustration showing reduction of:

* N_2
* CO
* hydrocarbons

by alternative nitrogenase systems.

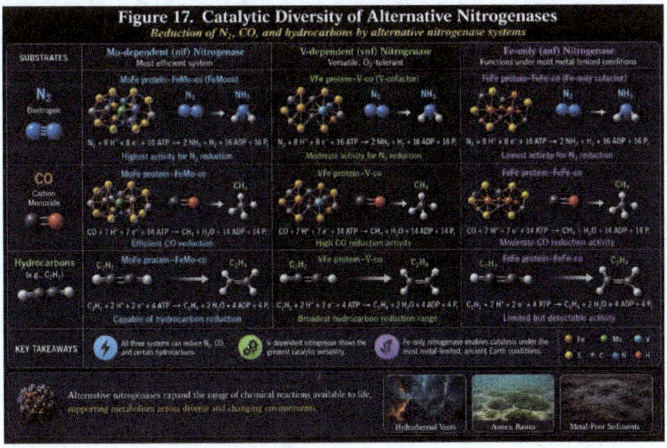

Figure 18

Nitrogenase Superfamily

Evolutionary overview linking nitrogenase to related metalloprotein systems.

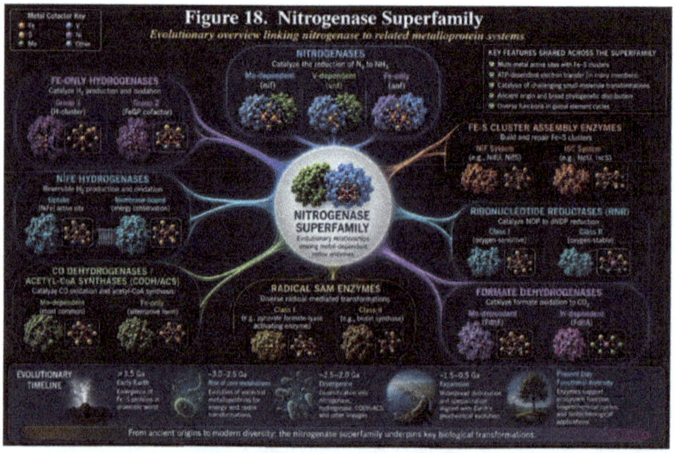

Figure 19

DPOR Structure and Function

ATP-dependent protochlorophyllide reduction mediated by DPOR.

Evolutionary relationship to nitrogenase highlighted.

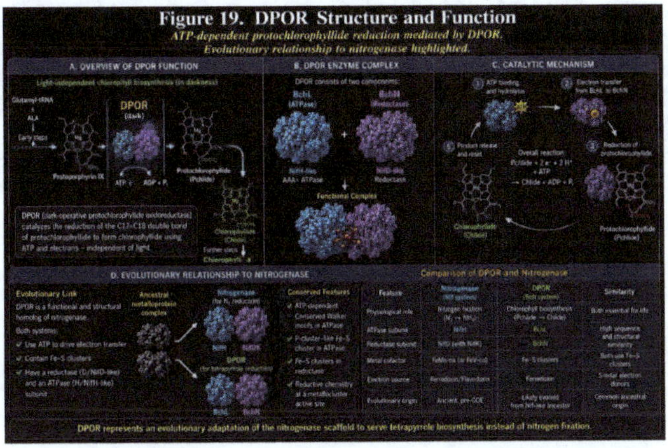

Figure 20

COR-Mediated Bacteriochlorophyll Biosynthesis

Role of COR in bacteriochlorophyll formation and photosynthetic diversification.

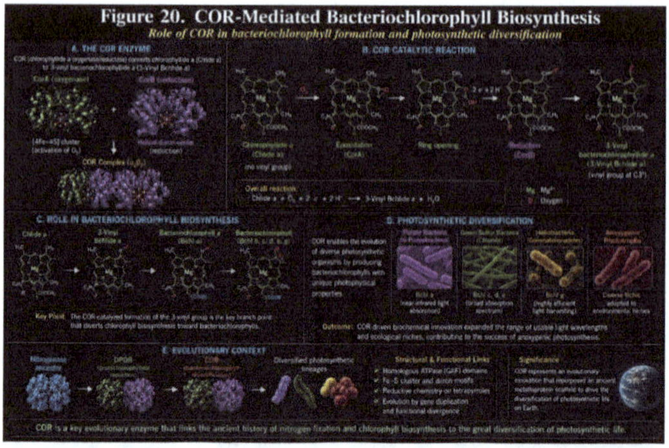

Figure 21

LPOR and Light-Driven Catalysis

Light-dependent chlorophyll biosynthesis mediated by LPOR using protochlorophyllide and NADPH.

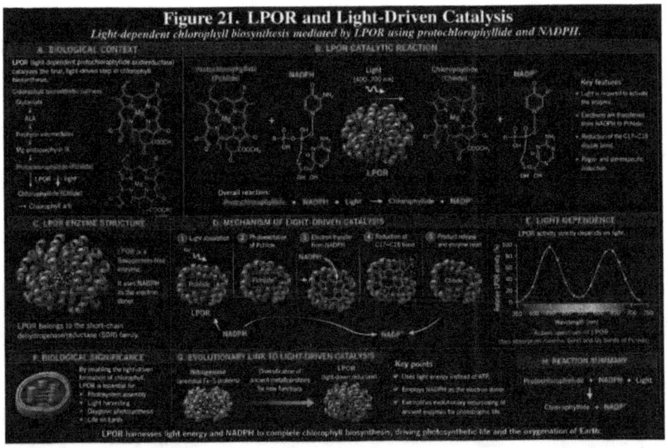

Figure 21. LPOR and Light-Driven Catalysis
Light-dependent chlorophyll biosynthesis mediated by LPOR using protochlorophyllide and NADPH.

Figure 22

Evolutionary Transition:

Nitrogenase → DPOR → COR → LPOR

Central evolutionary pathway linking nitrogen fixation and photosynthesis.

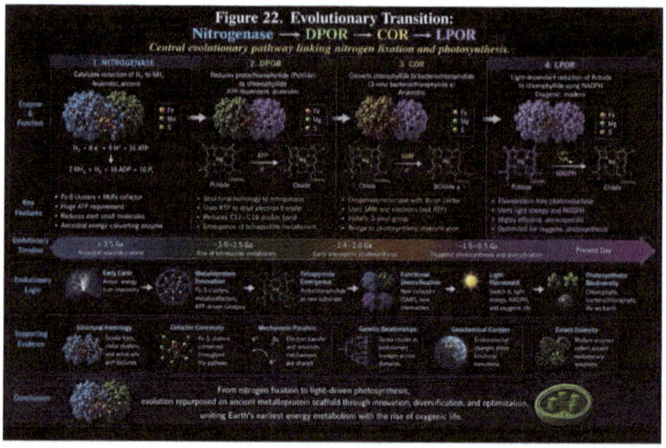

Figure 23

Origin of Photosynthesis

Proposed evolutionary progression from primitive phototrophy to oxygenic photosynthesis.

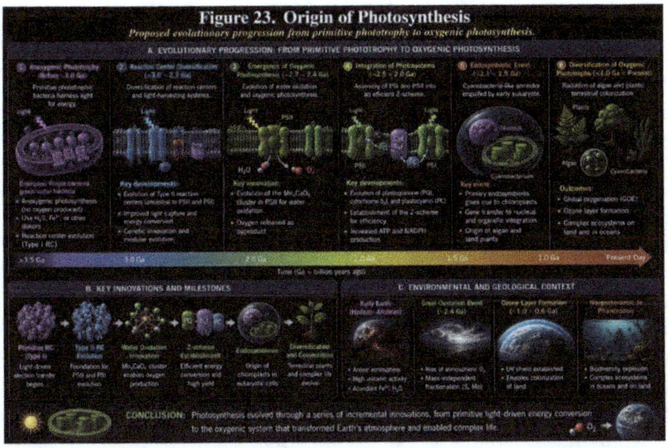

Figure 24

Cyanobacteria and Oxygen Production

Illustration of oxygenic photosynthesis and planetary oxygen release.

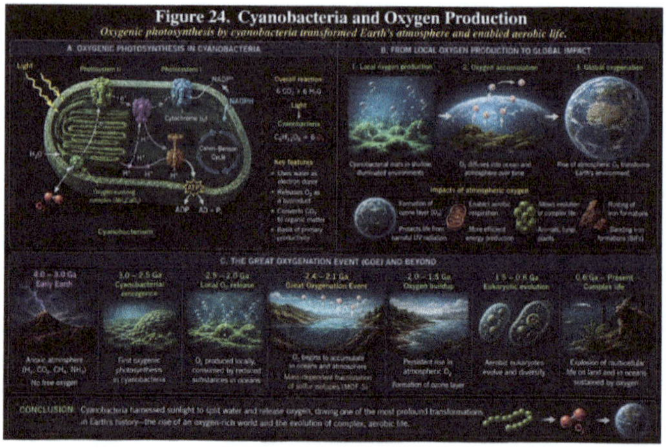

Figure 25

The Great Oxidation Event

Geological timeline showing atmospheric oxygen accumulation approximately 2.4 billion years ago.

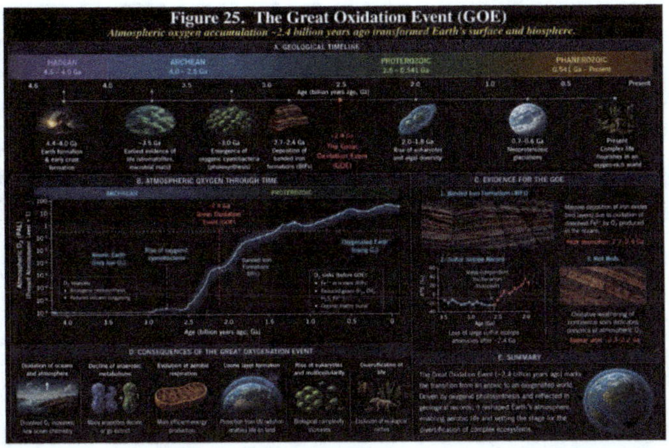

Figure 25. The Great Oxidation Event (GOE)
Atmospheric oxygen accumulation ~2.4 billion years ago transformed Earth's surface and biosphere.

Figure 26

The Oxygen Paradox

Conceptual diagram showing oxygen as both:

* Enabler of complexity
* Threat to nitrogenase

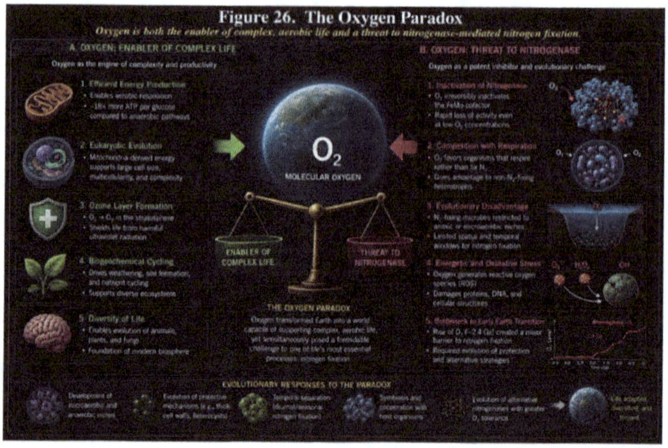

Figure 26. The Oxygen Paradox
Oxygen is both the enabler of complex, aerobic life and a threat to nitrogenase-mediated nitrogen fixation.

Figure 27

Heterocysts

Specialized cyanobacterial cells protecting nitrogenase from oxygen exposure.

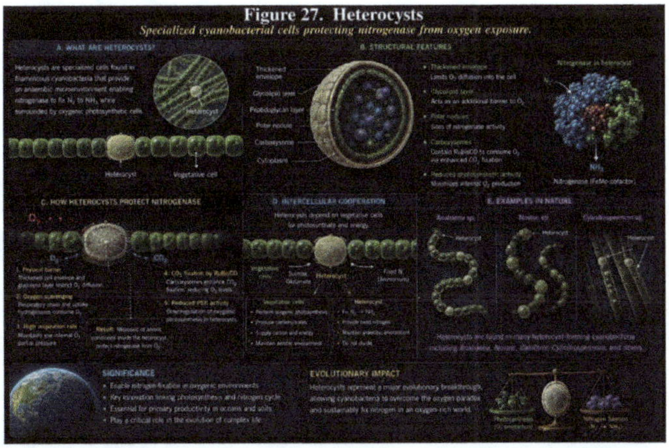

Figure 27. Heterocysts
Specialized cyanobacterial cells protecting nitrogenase from oxygen exposure.

Figure 28

Root Nodule Symbiosis

Interaction between legumes and rhizobia enabling biological nitrogen fixation.

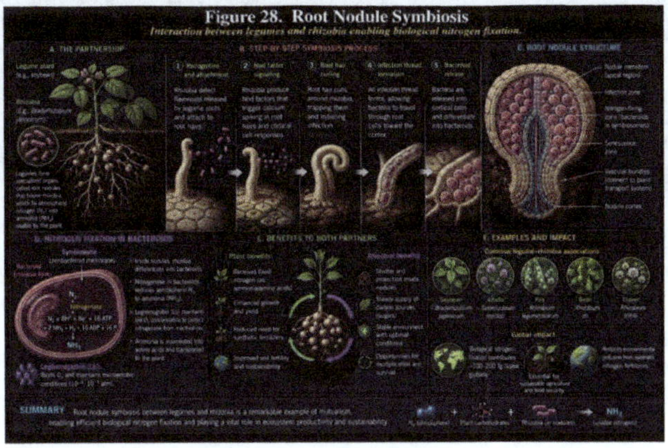

Figure 28. Root Nodule Symbiosis
Interaction between legumes and rhizobia enabling biological nitrogen fixation.

Figure 29

The Dual Engine Hypothesis

Photosynthesis and nitrogen fixation functioning as the two fundamental biochemical engines of the biosphere.

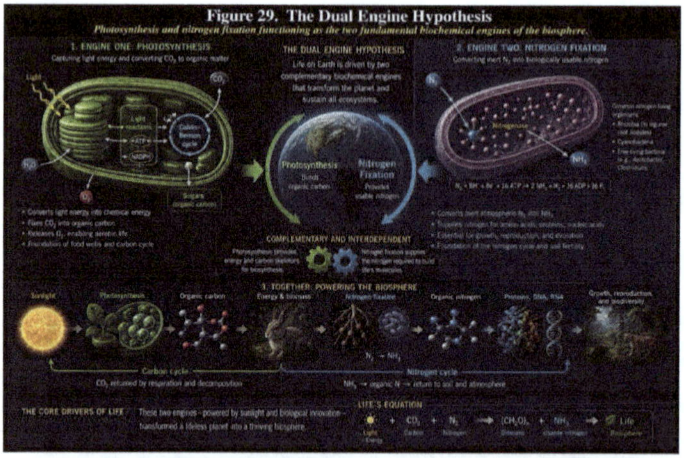

Figure 30

Nitrogen Threshold Hypothesis

Model showing how increasing fixed nitrogen availability enables increasing biological complexity.

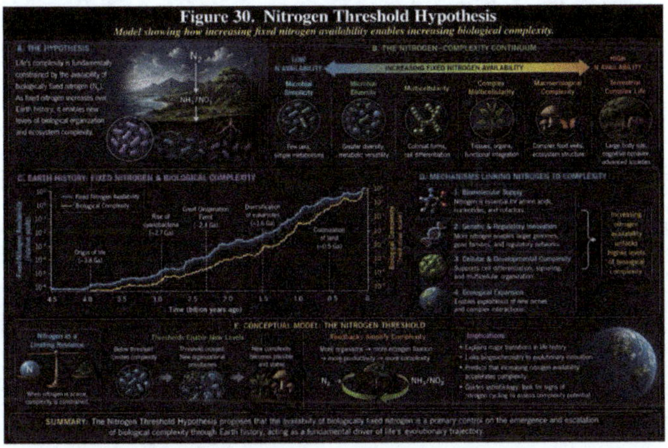

Figure 31

Nitrogen and Genome Expansion

Relationship between nitrogen availability and increasing biological information capacity.

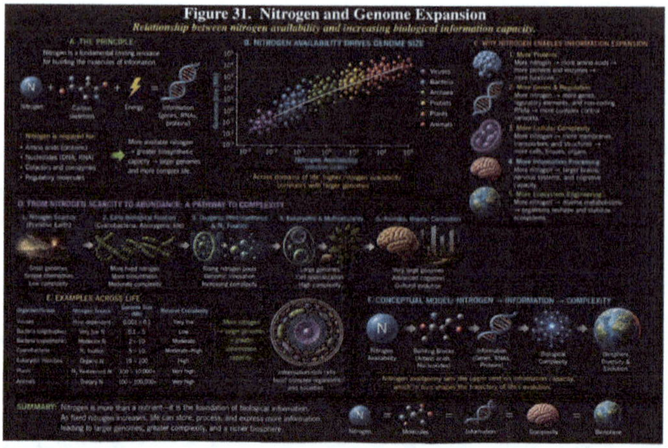

Figure 32

Nitrogen and Multicellularity

Role of fixed nitrogen in supporting multicellular evolution.

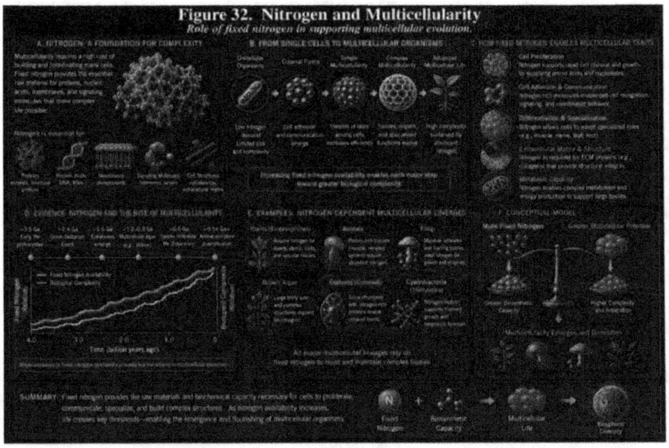

Figure 33

Nitrogen Flow Through Food Webs

Movement of biologically fixed nitrogen through plants, herbivores, carnivores and decomposers.

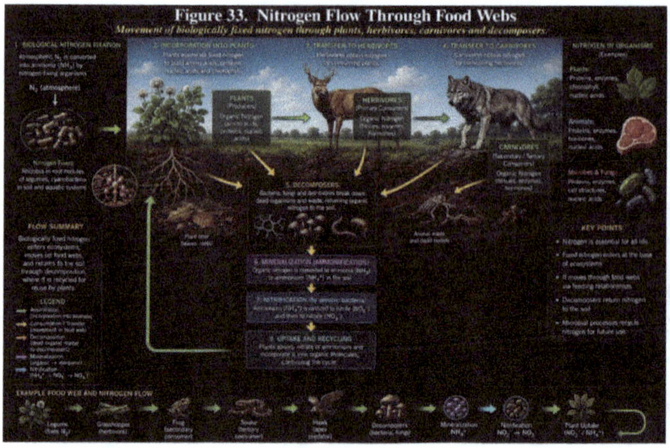

Figure 34

Engineering Nitrogen Fixation

Overview of modern strategies including:

* nif transfer
* symbiotic engineering
* chloroplast engineering
* synthetic biology

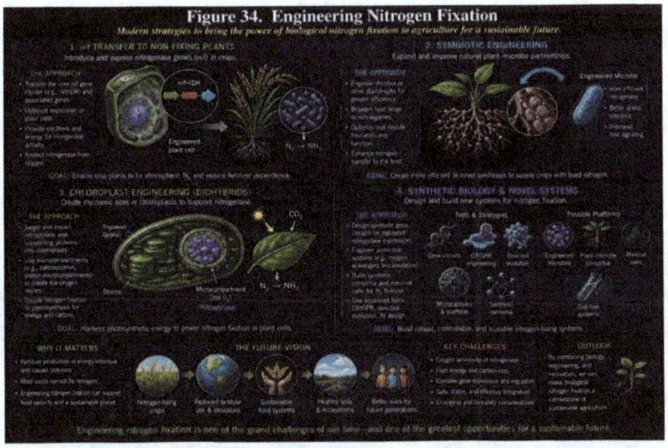

Figure 35

Nitrogen-Fixing Cereals

Conceptual illustration of future self-fertilising wheat, rice and maize.

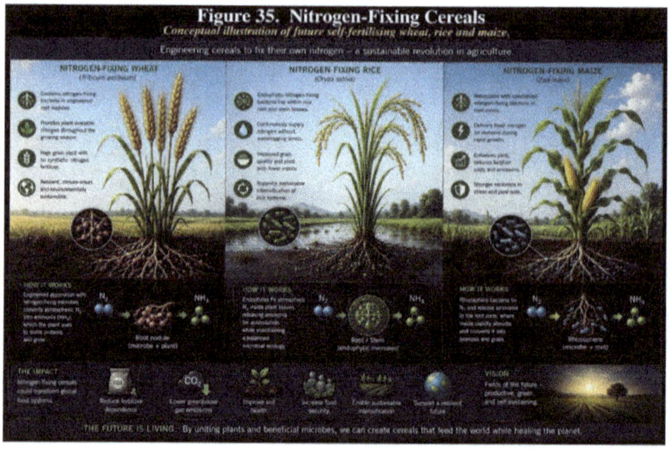

Figure 36

Chloroplast Engineering Platforms

Potential use of chloroplasts for engineered nitrogen fixation.

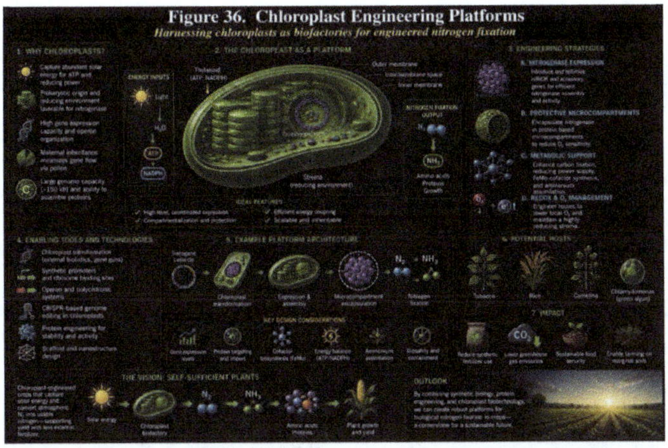

Figure 37

Synthetic Biology Design Cycle

AI-assisted Design → Construction → Testing → Optimization workflow.

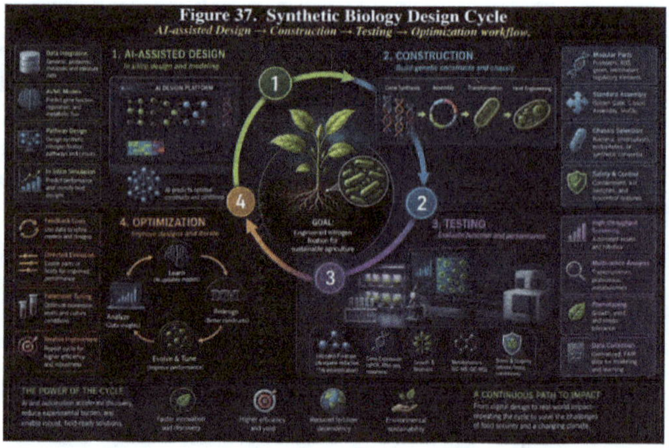

Figure 38

Synthetic Nitroplast Concept

A proposed nitrogen-fixing organelle integrated within plant cells.

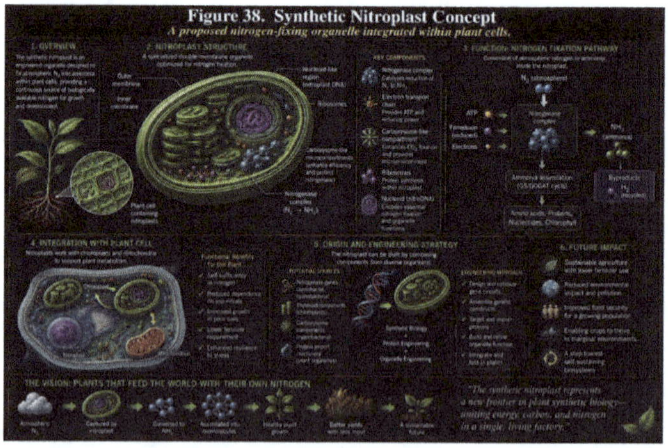

Figure 39

Evolution Toward Nitroplasts

Free-living diazotroph → symbiosis → intracellular residence → organelle.

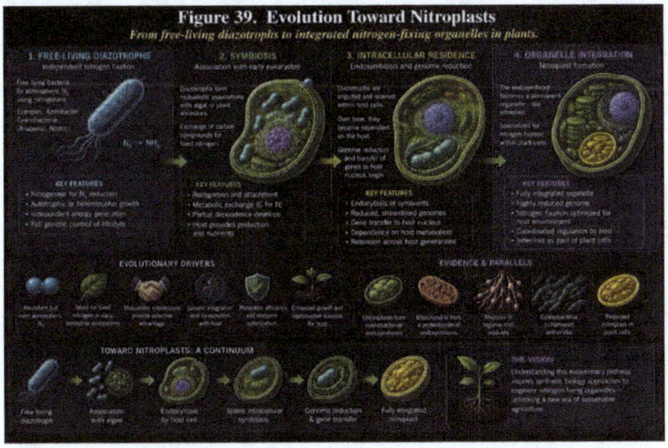

Figure 40

Nitroplast Revolution

Potential transformation of agriculture through synthetic nitrogen-fixing organelles.

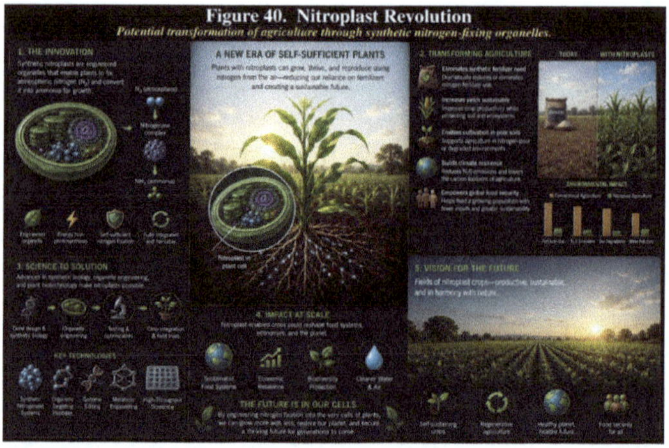

Figure 41

Light-Utilising Nitrogenase (LUN)

Conceptual framework showing direct coupling of photochemical energy to nitrogen fixation.

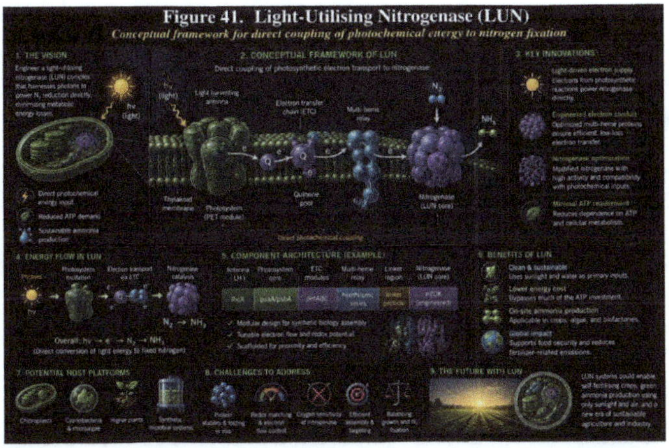

Figure 42

DPOR–LPOR–LUN Evolutionary Model

Proposed evolutionary pathway leading toward photochemical nitrogen fixation.

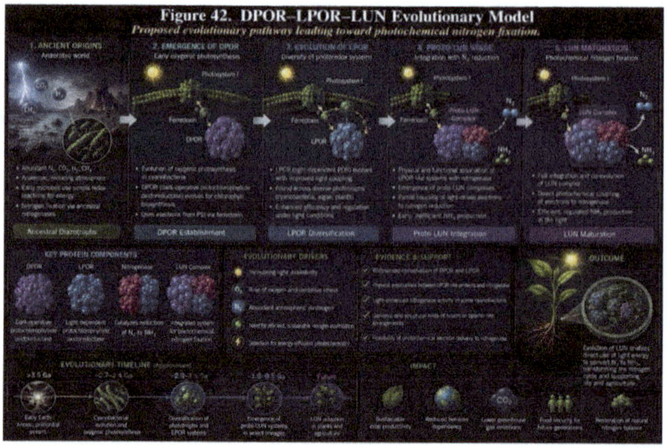

Figure 43

AI-Guided LUN Design

Integration of protein design tools, structural prediction and synthetic biology.

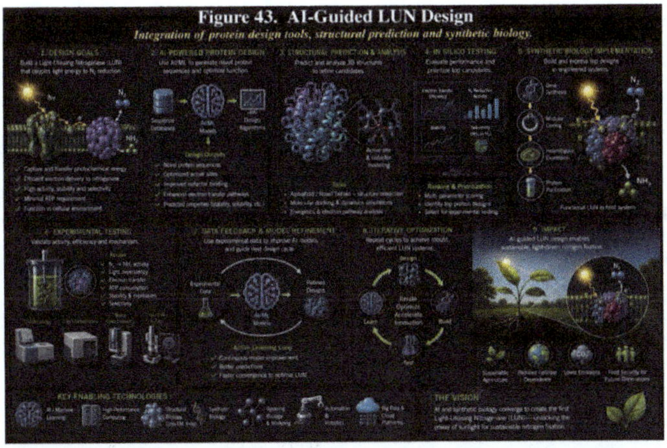

Figure 44

AI-Guided Evolution

Transition from natural evolution toward intelligent evolution.

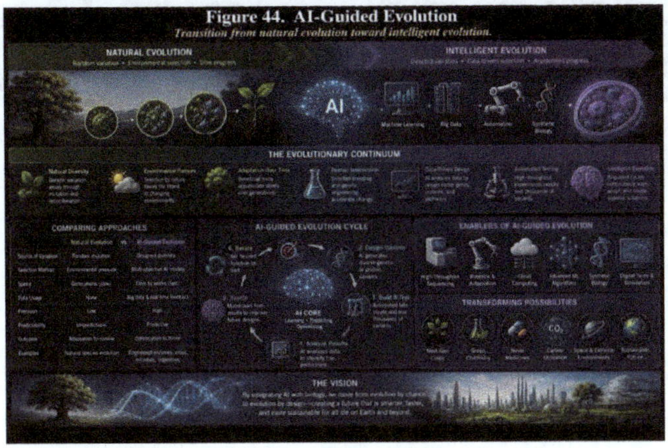

Figure 45

Self-Fertilising Crops

Future agricultural systems integrating biological nitrogen fixation directly into crop metabolism.

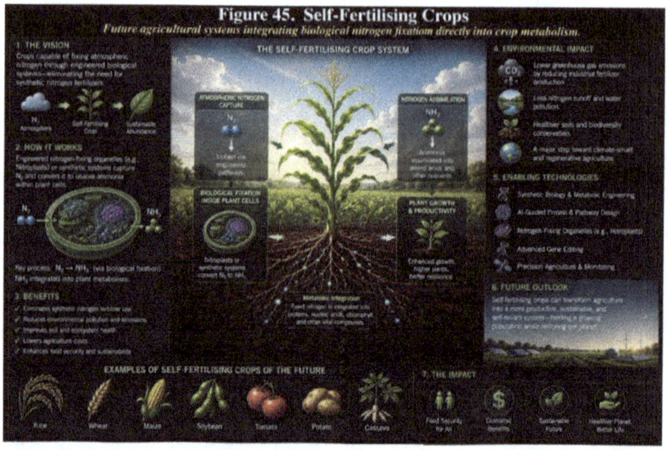

Figure 46

Space Agriculture Systems

Closed ecological systems supporting human life beyond Earth.

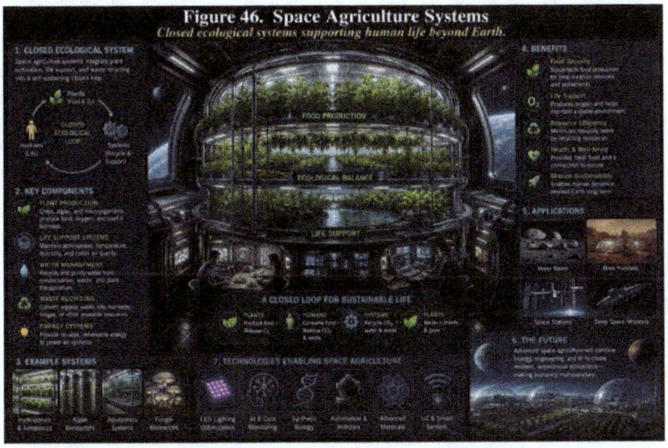

Figure 47

Nitrogen Fixation on Mars

Conceptual model of biological nitrogen cycling within Martian agricultural habitats.

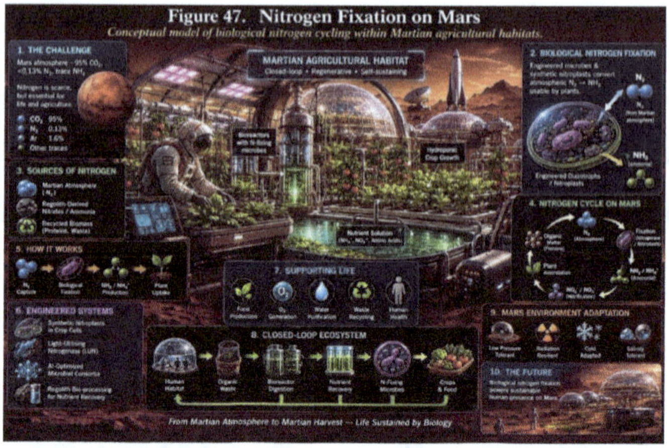

Figure 48

Cosmic Agriculture Theory

Expansion of agriculture from Earth to interplanetary ecosystems.

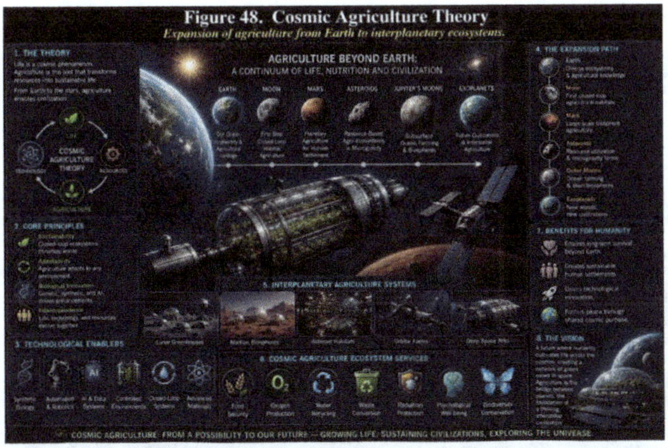

Figure 49

Earth Clock Unified Theory of Life

Integrated framework connecting:

* Nitrogen Fixation
* Photosynthesis
* Oxygenation
* Complexity
* Agriculture

* Artificial Intelligence

* Space Expansion

within planetary time.

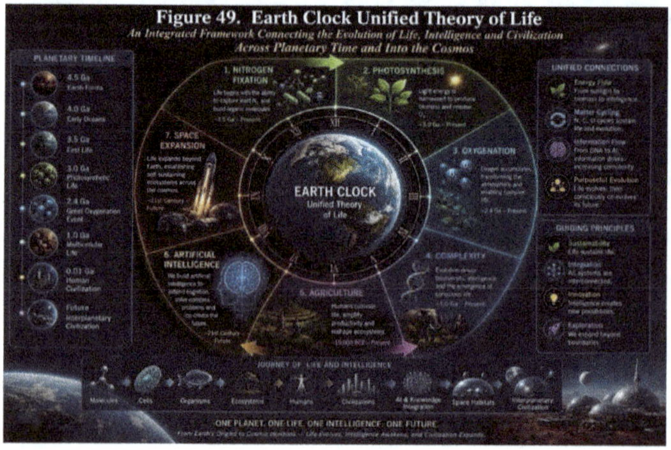

Figure 49. Earth Clock Unified Theory of Life

Figure 50

The Earth Clock

Final synthesis figure depicting Earth as a biological clock.

The diagram integrates:

* Origin of Life
* Nitrogen Fixation
* Photosynthesis
* Complex Life
* Humanity
* Artificial Intelligence
* Future Evolution

across Earth's complete habitable lifespan.

APPENDIX B

GLOSSARY OF TERMS

A

Acetylene Reduction Assay (ARA)

A classical method used to estimate nitrogenase activity by measuring the reduction of acetylene (C_2H_2) to ethylene (C_2H_4).

Ammonia Assimilation

The incorporation of NH_3 into amino acids and other nitrogen-containing biomolecules.

Anammox

Anaerobic ammonium oxidation.

A microbial process converting ammonium and nitrite into dinitrogen gas.

Anthropocene

The proposed geological epoch characterized by significant human influence on Earth's systems.

B

Biological Nitrogen Fixation (BNF)

The enzymatic conversion of atmospheric nitrogen (N_2) into ammonia (NH_3) by diazotrophic microorganisms.

Biosphere

The global ecological system integrating all living organisms and their interactions with Earth's atmosphere, hydrosphere and lithosphere.

C

Chlorophyll

The primary pigment responsible for photosynthetic light harvesting.

Cyanobacteria

Oxygen-producing photosynthetic prokaryotes that played a central role in the Great Oxidation Event.

D

Diazotroph

An organism capable of fixing atmospheric nitrogen.

DPOR

Dark-Operative Protochlorophyllide Oxidoreductase.

A nitrogenase-related enzyme involved in chlorophyll biosynthesis.

E

Earth Clock

The conceptual framework developed in this book.

Earth is viewed as a finite-duration biological system spanning approximately ten billion years of habitability.

- Fe Protein

- FeMoco
- Nitrogenase
- LPOR
- COR
- LUN
- Nitroplast
- Great Oxidation Event
- Oxygen Paradox
- Synthetic Biology
- Artificial Intelligence
- Planetary Evolution
- Complexity
- Exobiology
- Terraforming
- Space Agriculture

APPENDIX C

ABBREVIATIONS

Abbreviation	Definition
AI	Artificial Intelligence
ARA	Acetylene Reduction Assay
ATP	Adenosine Triphosphate
BNF	Biological Nitrogen Fixation
COR	Chlorophyllide Oxidoreductase
DPOR	Dark-Operative Protochlorophyllide Oxidoreductase
ETC	Electron Transport Chain
FeMoco	Iron-Molybdenum Cofactor
GOE	Great Oxidation Event
GS-GOGAT	Glutamine Synthetase–Glutamate Synthase Pathway
LUN	Light-Utilising Nitrogenase

Abbreviation	**Definition**
LPOR	Light-Dependent Protochlorophyllide Oxidoreductase
NADPH	Nicotinamide Adenine Dinucleotide Phosphate
nif	Nitrogen Fixation Gene Cluster
NH_3	Ammonia
N_2	Dinitrogen
PSI	Photosystem I
PSII	Photosystem II
ROS	Reactive Oxygen Species
Rubisco	Ribulose-1,5-bisphosphate Carboxylase/Oxygenase

……

APPENDIX D

THE EARTH CLOCK TIMELINE

Earth History and Future Evolution

Formation and Early Earth

Time (Ga)	Event
4.60	Earth forms
4.40	Stable oceans
4.10	Prebiotic chemistry
3.80	Earliest evidence of life
3.50	Biological nitrogen fixation
3.00	Photosynthesis expands
2.70	Cyanobacteria dominate
2.40	Great Oxidation Event

Rise of Complexity

Time (Ga)	Event
2.10	Eukaryotes
1.60	Multicellularity
1.00	Complex ecosystems
0.54	Cambrian Explosion
0.45	Land plants
0.25	Dinosaurs
0.066	Mammalian radiation
0.0003	Homo sapiens

Human Epoch

Time	Event
10,000 BCE	Agriculture
1800 CE	Industrial Revolution
1950 CE	Green Revolution
2025 CE	AI Revolution

Time	Event
2050 CE	Synthetic Nitrogen Fixation?
2100 CE	Self-Fertilising Crops?

Future Earth

Future Time	Event
+0.01 Ga	Advanced Synthetic Biology
+0.10 Ga	Global Planetary Engineering
+1.0 Ga	Oceans Begin to Evaporate
+2.0 Ga	Water Loss
+3.0 Ga	Surface >200°C
+4.5 Ga	Red Giant Sun
+5.0 Ga	End of Earth's Habitability

APPENDIX E

REFERENCES

Journal Articles

James A. Howard & Rees, D.C. (1996).

Structural basis of biological nitrogen fixation.

Chemical Reviews **96**, 2965–2982.

Nature Papers

Raymond S. Turner et al. (2024).

A nitrogen-fixing organelle in a marine alga.

Science **384**, 123–129.

Zhang S., et al., Cheng Q‡, Scrutton N‡. (2019). Structural basis for light-driven enzyme catalysis. Nature 574:722–725.

Books

Nick Lane (2015).

The Vital Question.

W.W. Norton, New York.

Classic Nitrogenase Literature

- Burris (1972)
- Dixon & Kahn (2004)
- Howard & Rees (1996)
- Raymond et al. (2004)
- Cheng (2008)
- Boyd et al. (2011)
- Mus et al. (2019)

Photosynthesis

- Blankenship (2014)
- Nelson & Ben-Shem (2004)
- Cardona (2019)

Earth Evolution

- Knoll (2015)
- Catling & Kasting (2017)
- Hazen (2022)

Synthetic Biology

- Keasling (2010)
- Khalil & Collins (2010)
- Cheng & Silver (2022)

Artificial Intelligence

- LeCun et al. (2015)
- Jumper et al. (2021)
- Baek et al. (2021)

Space Agriculture

- Wheeler (2010)
- Massa et al. (2016)
- NASA CELSS Program Reports

About the Author

Qi Cheng

Qi Cheng is a scientist, author, and interdisciplinary researcher whose work spans biological nitrogen fixation, photosynthesis, evolutionary biology, synthetic biology, and future agricultural systems.

He received his PhD from the University of East Anglia (United Kingdom) and has conducted research at several internationally recognized institutions, including the University of Cambridge. His scientific interests focus on understanding how fundamental biological processes have shaped the evolution of life on Earth across geological timescales.

Over the course of his career, he has worked on biological nitrogen fixation, nitrogenase evolution, cyanobacterial physiology, chlorophyll biosynthesis, algal biotechnology, and protein engineering. More recently, his research has expanded into synthetic biology, artificial intelligence-assisted protein design, engineered nitrogen fixation, synthetic nitroplast systems, and the concept of Light-Utilising Nitrogenase (LUN).

Qi Cheng is the creator of the Earth Clock framework, an interdisciplinary theory that views Earth as a finite-duration biological system operating across approximately ten billion years of planetary habitability. The Earth Clock integrates biological evolution, photosynthesis, nitrogen fixation, planetary oxygenation, ecological complexity, agriculture, technological development, artificial intelligence, and future space expansion into a unified temporal framework.

In addition to scientific research, he is the author of the Earth Clock Library, a long-term series exploring the history, present state, and future trajectory of life, civilization, and planetary evolution. Through this series, he seeks to connect deep-time Earth history with humanity's future development on Earth and beyond.

His current interests include:

• Biological Nitrogen Fixation
• Nitrogenase and Nitrogenase-like Enzymes
• Photosynthesis and Chlorophyll Evolution
• Synthetic Biology and Protein Engineering
• Artificial Intelligence for Biological Design
• Nitroplast Engineering
• Light-Utilising Nitrogenase (LUN)
• Future Agriculture and Food Security
• Earth System Evolution
• Space Agriculture and Interplanetary Civilization

Through research, writing, and scientific exploration, Qi Cheng advocates a long-term perspective on life and civilization, emphasizing the importance of understanding Earth's biological history in shaping humanity's future.

Author of the *Earth Clock Library*

University of Cambridge (Visiting Scholar/Fellow)

Henan University/Hebei Agricultural University

United Kingdom / China

www.ingramcontent.com/pod-product-compliance
Lightning Source LLC
LaVergne TN
LVHW021738071126
851857LV00040B/1570
9781919055008